# De Rebus Bellicis

## Part 1

Aspects of the *De Rebus Bellicis*

Papers presented to
Professor E. A. Thompson

Edited by M. W. C. Hassall

## Part 2

*de rebus bellicis*: the text
edited by Robert Ireland

B A R International Series 63
1979

B.A.R.

B.A.R., 122 Banbury Road, Oxford OX2 7BP, England



B.A.R.—S63, 1979: "De Rebus Bellicis: Part 1: Aspects of the De Rebus Bellicis; Part 2: the edited text"



ISBN 9780860540632 paperback

ISBN 9781407349473 e-book

DOI https://doi.org/10.30861/9780860540632

A catalogue record for this book is available from the British Library

This book is available at www.barpublishing.com

# INTRODUCTION

The Anonymus author of the tract known as the de Rebus Bellicis lived in the middle of the fourth century A.D., at a time when, in the words of a contemporary, "the most savage peoples were rousing themselves and pouring across the nearest frontiers as if trumpets were sounding the war-note throughout the whole Roman world". (Amm. xxvi.4.5.). It was a period which seemed to the Anonymus, no less than to historians of a later age, to represent a decline from Rome's pristine glory, a decline which led, as with the hindsight of history we know, to the sack of the City itself within a couple of generations and the fall of the Western Empire after barely a couple more. Yet the Eastern Empire survied for centuries and the reasons for the collapse of the West did not lie in greater moral degeneracy or economic incompetence, but rather in a demographic process that had its beginning centuries before the Anonymus wrote.

When Caesar checked the westward migration of the Helvetii during the first century B.C. and forced Ariovistus and his Germans to return across the Rhine, he set the course of history back for four hundred years, but he could not change it; sooner or later the over-extended land frontier of the Western Empire was bound to fail. That it failed at last on Christmas Day 406 A.D. when the Rhine froze and enabled the waiting barbarians on the other side to cross, is not in itself surprising: the remarkable fact is that it did not fail sooner. The fatal weakness of the river frontiers of Rhine and Danube had been revealed with brutal and sudden clarity in the second century A.D. when the Marcomanni had crossed the Upper Danube and swept into Northern Italy only to be driven out at great cost and with utmost difficulty by the philosopher emperor Marcus Aurelius. The potential danger, however, had been recognized already even earlier, in the first century when Tacitus in the Germania had warned Rome of the peril she would face, if the feuding German tribes should stop fighting amongst themselves and only combine against her. The confederations of tribal groups like the Alamanni that emerged in the third century were the realization of the very threat that the historian had forseen.

In the third century A.D. the pressure built up to breaking point, not only on both river frontiers but in the East too, where the new Sassanid dynasty converted the quiescent and moribund Parthian Empire into a formidable and hostile neighbour. But the empire did not fall. Its survival was due to the exertions of a succession of strong but short lived warrior-emperors drawn from the Danubian provinces. The last and greatest of these men, Diocletian, was also an administrator of genius. He recognized the fact, not so obvious at the time, that the empire's vulnerability lay not only, or even primarily, in military matters but in her own internal political and economic weakness. Only from economic strength could Rome afford an increased army and only with political stability could she risk it. The reforms undertaken by Diocletian gave the Empire a new lease of life even though in the short term some of his most famous political and economic measures failed. Paradoxically the social, economic and military changes brought about by Diocletian and his successor Constantine were almost too successful. The creation of a strong and at first efficient administrative machine coupled with that of a large and militarily effective army bound by ties of loyalty to the house of Constantine, gave the empire both internal stability and relative peace on the frontiers. But the economic base proved too small to support both army and bureaucracy and those best able to bear the burdens of increased taxation, the great magnates were the most able to avoid their obligations.

It is against this background that the composition of the work known to scholars as the de Rebus Bellicis must be set. The title itself is not the original one which is unknown. Professor E. A. Thompson, points out that it is in fact misleading since the little treatise deals not merely with military matters but with 'defence from strength' in the sort of terms in which - if it is possible to make such a comparison - the reforms of Diocletian were conceived. Modest though they are, the Anonymus' proposals cover a whole spectrum from monetary and economic reform to military recruitment and the building of forts. His military proposals are however made the more striking by the inclusion of sections on war machines, and it is for this reason no doubt that the title de Rebus Bellicis was later coined. The machines themselves range from the purely fanciful, such as the scythed chariots, and most famous of all, the ox-powered paddle boat, to weapons and equipment actually in use by the armies of the fourth century, such as the weighted javelines, martiobarbuti or plumbatae. In some instances it is not clear whether he is describing original inventions of his own or pieces of existing military equipment.

A unique feature of the de Rebus Bellicis and one that has always enhanced its interest, is the inclusion of illustrations. Even in the medieval versions of the lost Carolingian copies of the originals, it can be seen that they often make clear points that are obscure in the Anonymus' text. It is certain that he relied on them to make his meaning clear for he refers to them on a number of occasions and sometimes in terms of obvious pride (e.g. VI.5. VII.1). An example of the superiority of illustration over text descriptions occurs in the section on the Ballista Fulminalis, with its two pairs of pulleys (not mentioned in the text, but clear in the accompanying picture), one to pull back the "slider" on which the arrow rests, and the second to pull it forward again after discharge. It is this accuracy of detail when the text is at its most unhelpful and obscure that raises the question whether the artist was not copying actual machines whose method of working neither he nor our Anonymus author understood. Yet in attempting to interpret the pictures of the machines and decide whether they could actually have worked, it is necessary to compare the different illustrations in the principal MSS of the de Rebus Bellicis. For the first time these are now published in Part 2 together with the line blocks of the seventeenth century editio princeps which have been inserted in their correct position in the Latin text. The most striking thing about the illustrations is the difference in style adoped by the various copyists; in one case, the illustrations of the Anonymus' proposed new coin types, one of the medieval copyists has used actual examples of Roman coins.

The Anonymus was perhaps unique among contemporaries in his attempt to analyse the causes of the crisis that had befallen the Empire and suggest specific cures. But could actual economic and military reforms on the scale suggested ever have been carried out, and would they have succeeded in postponing the collapse of the Western Empire? There is no doubt that strong economic and military reforms not only could have been put into effect, but were actually carried out in the Anonymus' own life time, and that in the short term they were beneficial. In 356 Julian, administering the prefecture of the Gauls for his uncle Constantius II both reduced the tax rate to 28% of its former value, and even succeeded in getting more money for the state than previously, party by overhauling the corrupt and inefficient machinery for collecting the amount due, and partly because the sum demanded was more realistic and tax payers did not attempt to avoid payment. As a result the tax payers "as if clear sunshine had beamed upon them after an ugly darkness, expressed their joy in gaiety and dances" (Amm. xv 5.14). Probably of longer lasting effect were the military reforms of Valens and Valentinian, which revitalised the defences of the empire during the period in which the Anonymus wrote. But a greater military effort meant a greater contribution from the economy to support it, and this the Empire, in the fact of a desperate shortage of man-power, was in no position to make. It was for this reason that the Anonymus concerned himself with the "mechanization" of the army and, time and again, he emphasizes how his schemes will result in a more economical use of the stretched resources of labour available. If some of his proposals seem naive, there is more to many of them than may appear at first sight. It is the object of several of the papers in Part 1 to assess how effective they would in fact have been. But, effective or not, at least the Anonymus deserved the credit for bravely attempting an analysis of the problems that beset the empire and for suggesting rational solutions to them.

It is a gentle irony that the Anonymus, whose words may never have been read by the emperors to whom his little treatise is addressed, should have attained to a certain posthumous fame, first of all in the middle ages when his machines in particular, caught the fancy of students of warfare and engineering, and subsequently in the period from the Renaissance to the present when his views on the crisis of the fourth century became of interest to historians of the later Roman Empire. Current interest in the Anonymus in Britain is due partly to the more generous space in the syllabuses of British Universities allotted to the late empire and partly due to the availability of Professor Thompson's excellent translation and commentary published in 1952, precisely four hundred years after the appearance of the first printed edition by Gelenius.

The papers that are here collected were given at a small seminar on the Anonymus held at the Institute of Archaeology of London University in December 1973. There is not one of those who contributed who has not benefitted immensely from Professor Thompson's scholarly edition of the text, translation and commentary, and, in offering these papers to him we are happy to acknowledge our debt. His commentary has, in almost every case, acted as the starting point for our detailed discussions of the different aspects of the _de Rebus Bellicis_, and his sympathetic approach to the Anonymus himself has provided a lead that many of us are happy to follow. Professor Thompson's _A Roman Reformer and Inventor_ is unhappily now out of print, and partly for this reason and partly to make the present collection more self contained, a newly revised text and a translation have been prepared by Robert Ireland (Part 2): the text often reflects Professor Thompson's recension, as all texts subsequent to his must (to a greater or lesser extent) while the editor has based his reports of the MSS on Professor Thompson's collations, and we acknowledge the kind permission of the Delegates of the Clarendon Press to make use of the Oxford edition in this respect. Beyond this, the text aims at independence; the translation was made without reference to Professor Thompson's version; and the notes on text, language and style attempt to consider the treatise fully, from these points of view, for the first time.

The papers are reproduced in Part 1 in the form in which they were given though in certain cases authors have carried out some revision in the light of discussion at the seminar. The late C. E. Stevens' summing up has been transcribed from a tape recording made at the time, with the minimum of editorial interference. Finally the fact that these papers should not only appear at all but at such an economical price is due to the post-war technical innovations in printing, adopted by B.A.R. - a state of affairs in which the Anonymus would surely have derived profound satisfaction!

# ASPECTS OF THE DE REBUS BELLICIS

Papers presented to Professor E. A. Thompson
at a seminar held in London, December 14 to 15, 1973

by

A. Cameron, J. J. G. Alexander, M. E. Henig, K. D. White,
P. Vanags, R. M. Reece, J. S. Johnson,
M. W. C. Hassall, P. A. Barker, D. H. Sherlock, J. P. Wild,
D. H. Hill, H. W. M. Hodges and C. E. Stevens

edited by

M. W. C. Hassall

CONTENTS

LIST OF PLATES

## LIST OF TEXT FIGURES

## ACKNOWLEDGEMENTS FOR ILLUSTRATIONS

Plate XII is reproduced by gracious permission of Her Majesty Queen Elizabeth II. Grateful thanks are also due to the following for permission to reproduce photographs:

A. C. L. Bruxelles: Pls. Xa, Xb.

Biblioteca Apostolica Vaticana: Pls. I, IIa, IIb, IV, V, VI, VIIIb.

Bodleian Library, Oxford: Pl. VII.

British Museum: Pl. VIIIa, Pl. XIII.

Warburg Institute, London: Pl. IX.

## ABBREVIATIONS

Where possible abbreviations are those employed by Jones, Later Roman Empire III, 392-406, to whom the reader is referred for published editions of ancient texts.

| | |
|---|---|
| AJP | American Journal of Philology |
| Amm. | Ammianus Marcellinus |
| Ant. Class. | L'Antiquité Classique |
| Ant. J. | Antiquaries Journal |
| Arch. Anz. | Archäologischer Anzeiger |
| B. Jb. | Bonner Jahrbücher |
| B.Vbl. | Bayerische Vorgeschichtsblätter |
| Byz. Zeitschr. | Byzantinische Zeitschrift |
| CEHE | Cambridge Economic History of Europe |
| CIL | Corpus Inscriptionum Latinarum |
| CJ | Codex Justinianus |
| CP | Classical Philology |
| CR | Classical Review |
| CTh | Codex Theodosianus |
| $CW^2$ | Transactions of the Cumberland and Westmorland Antiquarian and Archaeological Society. New Series. |
| DRB | De Rebus Bellicis |
| Dig. | Justinian, Digesta seu Pandectae |
| Downey | G. Downey et al. (eds.), Themistii Orationes quae supersunt (Teubner edition, Leipzig, 1965-74) |
| Econ. Hist. Rev. | Economic History Review |
| Ed. Diocl. | Edictum Diocletianum |
| IGBR | G. Mihailov (ed.), Inscriptiones Graecae in Bulgaria Repertae, Vols. I-IV (Sofia, 1956-66) |
| ILS | H. Dessau, Inscriptiones Latinae Selectae (Berlin, 1892-1916) |

| | |
|---|---|
| Isid. Etym. | Isidore Etymologiae sive Origines |
| JBAA[3] | Journal of the British Archaeological Association. Third series. |
| JRS | Journal of Roman Studies |
| Jh.OAI | Jahreshefte des Osterreichischen Archaeologischen Instituts in Wien |
| Joh. Lydus Mag. | Johannes Lydus, De magistratibus |
| Jones, Later Roman Empire | A.H.M. Jones, The Later Roman Empire, 284-602, a social, economic and administrative survey, I-III (Oxford, 1964) |
| Lewis and Short | C.T. Lewis and C. Short, A Latin Dictionary |
| Lib. Or. | Libanius, Orationes |
| Liddell-Scott-Jones | H.G. Liddell and R. Scott, A Greek-English Lexicon. Revised by Sir H.S. Jones (Oxford, 1968) |
| MGH | Monumenta Germaniae Historica |
| Malalas | John Malalas, Chronographia |
| KNKOB | Nieuws-bulletin van de Koninklijke Nederlandse Oudheidkundige Bond |
| Not. Dig. | Notitia Dignitatum |
| Num. Chron. | Numismatic Chronicle and Journal of the Numismatic Society |
| OMROL | Oudheidkundige Mededelingen uit het Rijksmuseum van Oudheden te Leiden |
| PLRE | A.H.M. Jones, J.R. Martindale and J. Morris, The prosopography of the later Roman Empire, Vol. I, A.D. 260-395 (Cambridge, 1971) |
| P. Oxy. | The Oxyrhynchus papyri (Egypt Exploration Society, Graeco-Roman memoirs) |
| P. Ryl. | A.S. Hunt et al (eds.), Catalogue of the Greek Papyri in the Rylands Library (Manchester, 1911, in progress) |
| Palladius, De Re Rust. | Palladius, De Re Restica |
| Pan. Lat. | Panegyrici Latini |
| Pfalz. Mus. | Pfalzisches Museum |
| Pollux, Onom. | Pollux, Onomasticon |
| Proc. Aed. | Procopius, De Aedificiis |
| Proc. HA | Procopius, Historia Arcana |

| | |
|---|---|
| PW(-K) | Pauly-Wissowa(-Kroll), *Real-Encyclopadie der classischen Altertumswissenschaft* |
| RCHM, *Eburacum* | Royal Commission on Historical Monuments, England, *The City of York: Vol I, Eburacum, Roman York* |
| *REA* | *Revue des Études Anciennes* |
| *RIC* | H. Mattingly *et al*, *Roman Imperial Coinage* (London, 1923, in progress) |
| *Rev. Arch.* | *Revue Archéologique* |
| *Rev. Phil.* | *Revue de Philogie* |
| *SHA* | *Scriptores Historiae Augustae* |
| Soc. *EH* | Socrates, *Historia Ecclesiastica* |
| Soz. *EH* | Sozomen, *Historia Ecclesiastica* |
| *TLL* | *Thesaurus Linguae Latinae* |
| Them *Or*. | Themistius, *Orationes* |
| Thompson | E.A. Thompson, *A Roman Reformer and Inventor* (Oxford, 1952) |
| Val. III, *Nov.* | Valentinian III, *Novellae* |
| Veg. | Vegetius, *Epitoma rei militaris* |
| *V. Severini* | Eugippius, *Vita S. Severini* |
| Vergilius Romanus | E. Rosenthal, *The Illuminations of the Vergilius Romanus* (Zürich, 1972) |
| Zos. | Zosimus, *Historia Nova* |

# THE DATE OF THE ANONYMUS *DE REBUS BELLICIS*

by Alan Cameron

*Constantini temporibus* at II.1 gives a firm *terminus post quem* of 337, the year of Constantine's death. And (as has generally been agreed) there are various indications that the work was written before Adrianople in 378. It is hardly necessary to quote more than VI.1.2.

> 'Above all it must be recognized that wild nations are pressing upon the Roman Empire and howling round about it everywhere, and treacherous barbarians, covered by natural positions, are assailing every frontier. For usually the aforesaid nations are either covered by forests or occupy commanding mountain positions or are defended by snow and ice ...'
>
> (Thompson's translation).

We may compare the urgent recommendation to build forts at intervals of a mile along all the frontiers at XX.1, and the boast that the *Ballista fulminalis* can shoot right across the Danube (XVIII.5). It is clear that the Anonymus wrote at a time when barbarians, however threatening, could at any rate be assumed to live *beyond* the imperial frontiers.

He seems also to have written at a time when war with Persia was no very distant prospect (XII.1, XIX.4); indeed, the references to the storming of city and fortress walls (VI.4, VIII.1) and to marching inside Persian territory seem to envisage a Roman invasion of Persia.

The work is addressed to a plurality of emperors (*pref.* 1, 9), who (on the natural interpretation) had a plurality of sons (*pref.* 8), and had suppressed a plurality of usurpers (II.6).

This is quite a lot to go on. It should not be too difficult to pin down a moment between 337 and 378 when a plurality of emperors with a plurality of sons suppressed a plurality of usurpers in the shadow of a Persian war.

It has long been seen[1] that only one period will answer all these conditions, the joint reign of Valentinian and Valens: some time between 366 (when Valens' son Valentinianus Galates was born) and 375 (when Valentinian died). It is (I believe) possible both to narrow the *termini* a little and to establish at least whether the work was written in the East or West. But first we must deal with Santo Mazzarino's elaborately argued case for the period 353-360, when Constantius II was Augustus and Gallus and Julian his successive Caesars.[2]

Mazzarino rightly points out that the body of the work is addressed to just one emperor (invicte imperator, XVIII.7; sacratissime imperator, XXI.1). This one emperor, he claims, must be the senior emperor. This is proved, he believes, by the closing appeal to the emperor to 'throw light upon the confused and contradictory rulings of the laws' (XXI). Only the senior emperor, he claims, was entitled to codify the law.

If this is right, then, Valentinian being senor to Valens, if the work was written under them, the addressee would have to be Valentinian. It would then follow that it is Valentinian who is credited with putting down the usurpers. This would mean that the most obvious candidates, namely the eastern usurpers Procopius and his kinsman Marcellus in 365-6, would be ruled out. For if the Anonymus had been addressing Valentinian, he could hardly have credited him with suppressing a rebellion which, after careful deliberation, he had left entirely in the hands of Valens[3]

The only usurpers that can be produced for Valentinian are the obscure Pannonian rebel Valentinus[4] and the African prince Firmus. Now while I fully agree with Mazzarino that Valentinus will not do, Firmus should hardly have been dismissed so cavalierly, even if ultimately he too proves unacceptable (see below). Mazzarino however, convinced that by eliminating Valentinian he had eliminated Valens too, found himself driven back to Constantius and his Caesars Gallus and Julian.

Here we do have a satisfactory pair of usurpers, Magnentius and Decentius (350-3), and fairly constant hostility from Persia - but (despite Mazzarino) not one of the other conditions is satisfied.

To start with, not one of the three had a son. Mazzarino argues that filius is being used in an extended sense, meaning in effect 'successor'. Now an emperor certainly could have described his junior colleagues and designated successors as his filii.[5] Yet even so Constantius and his Caesars simply do not fit our requirements. For it is not the one emperor addressed in the body of the work who is credited with sons (real or metaphorical), but the 'clementissimi principes' of pref. 8. Now Gallus and Julian were Caesars in succession (from 351-4 and 355-9 respectively). Thus at no time could anyone have addressed Constantius II and his one Caesar of the moment as 'most merciful emperors... who transmit to your sons the affection due to the Roman name' ('qui Romano nomini debitos affectus propagatis in filios'). The Caesar of the moment would have to be counted both among the 'merciful emperors' (to justify the plural) and as the 'sons' - all by himself!

Less formally conclusive, perhaps, but hardly less serious an objection is the sharp and open criticism of the fiscal policy of Constantine in II.1f. Is it really credible that a man whose difficult task was to persuade the suspicious and vindicative Constantius to give a sympathetic hearing to his drastic proposals for reform should have been so insanely tactless as to begin by denouncing Constantius' father as the author of the social and economic ills he was proposing to cure? Thanks to Richard Reece it can now be added that, despite his establishment of the solidus, gold was not 'assigned to petty transactions' (II.1) under Constantine. That did not come until Constantius. Nor could anyone writing under Constantius have made such an error.[6]

We must return, then, to Valentinian and Valens. But which? There is, I think, no justification whatever for Mazzarino's belief that the one emperor must be the senior emperor.[7] If Anonymus had been writing anywhere in the eastern provinces, let alone (as may well have been the case) at Valens' court, then surely he would have addressed his proposals to Valens rather than Valentinian - though he might well have thought it appropriate to address his more general prefatory remarks to both emperors.

As for the request for a law code, since each emperor could legislate, each must also have been able to codify the law. It is true that, if Valens had issued a code, it would not automatically have been accepted by Valentinian, but only for the same reason that individual laws issued by one emperor were not automatically valid in his colleague's territory. In practice one emperor normally did ratify his colleague's laws, and we may be sure that (for the reasons Anonymus gives) Valentinian would have welcomed a code. In the event the job was not done till the Theodosian Code of 438, which was accepted in the West - but because Valentinian III endorsed it[8] not because Theodosius II was the senior Augustus. The last two codes before that were issued by private citizens,[9] so there can have been no possible objection to an eastern writer putting the simple and reasonable request made in XXI to Valens.

If the one emperor is Valentinian, then all we have by way of usurpers for him is Firmus. Mazzarino, following here a widespread modern view, objected that Firmus 'non è vero e proprio _tyrannus_'[10] (p.385, n.66). The argument is that, as a Moor, Firmus was an 'external' rather than 'internal' enemy, an invader rather than a usurper.[11] Whether Firmus saw it this way may be doubted, since he took the trouble to have himself formally crowned with the military _torques_, just like Julian a decade earlier.[12] But Roman contemporaries do seem to have been in some doubt how to classify him, and for all we know Anonymus might have been among those who reckoned him a barbarian.

There are other objections, however. First and least substantial, Anonymus says _tyranni_ in the plural. Now while it would be simple enough to parallel the 'rhetorical plural', 'usurpers like Firmus' meaning in fact just Firmus, _prima facie_ more than one usurper is indicated.

Firmus' rebellion dragged on into 375. It was in November of that year that Valentinian died. So if it is the suppression of Firmus that Anonymus has in mind, his book could be dated to within a month or so in mid 375. Attractive though such precision would be, there are three objections to so late a date:

(1) The _prima facie_ implication of _pref._ 8 is that each emperor had a son, and (particularly in the light of the references to 'propagating' and 'affection') a real, not a metaphorical son. Now Valens' only son died in the course of 370.[13]

(2) _DRB_ XX advocates the building of frontier forts, a policy that both Valentinian and Valens had been pursuing, on the Rhine and Danube frontiers respectively, with great energy since 369. It is difficult to believe that Anonymus would have been either so foolish or so ill-informed as to come out with such a proposal as late as 375.

(3) According to II.5, 6 these _tyranni_ found their support among the poor, made desperate by overtaxation and inflation:

'the poor were driven by their afflictions into various criminal enterprises ... laying waste the fields, breaking the peace with outburst of brigandage, stirring up animosities; and, passing from one crime to another, supported usurpers, produced by poverty for your glorification rather than inspired by their own bravery.'

Now it was claimed at the time that Firmus' revolt was sparked off by the extortions of the villainous Romanus, count of Africa. Whether or not this was true, Romanus was subsequently arrested, and two high officials committed suicide on being implicated in an investigation set in motion by the energetic prefect Maximinus. Another death that seems to have had some connection with the activities of Maximinus is the sudden execution of Theodosius, the general responsible for Firmus' defeat. This of course proved most embarassing when his son became emperor in 379, and the facts were very thoroughly buried. Romanus himself survived the death of Valentinian, but a new trial was launched by the new emperor Gratian, the outcome of which is unknown.[14]

The 'Romanus affair' was a murky business, with ramifications that led dangerously close to the throne, a scandal on the grand scale. If Anonymus had been writing in 375, he would have been very ill advised to link the rebellion of Firmus so emphatically with official extortion.

All in all Firmus does not seem a very plausible candidate for Anonymus' *tyranni*. And with Firmus goes Valentinian's claim to be the principal addressee. That leaves Valens. *His* usurpers, Procopius and Marcellus, fit the bill to perfection.

To start with there were two of them, Marcellus taking up the standard after Procopius' defeat and death.[15] Thus their rebellions are closely linked, as must have been those Anonymus had in mind, if his *tyranni* all drew their support from among the victims of the same economic pressures.

Then there are the accounts of Procopius' supporters in Themistius and Ammianus (Themistius an exact contemporary, writing in 366/7; Ammianus writing about twenty years later). According to Ammianus they were drawn from 'dealers in cheap dainties' (*cuppediarum vilium mercatores*),[16] and, more to the point, 'some from the dregs of the people, led by desperation or by blind ambitions' (*ex vulgari faece nonnulli, desperatione consiliisque ductantibus caecis*),[17] and (closest of all to Anonymus' words), 'a few, whom their worthlessness and desperation drove to crime' (*pauci, quos vilitas et desperatie trudebat in scelus*).[18] Themistius, in a speech purporting to congratulate Valens on his generous treatment of Procopius' followers, describes him as winning them with promises of 'abolition of debts and redistribution of land' χρεῶν ἀποκοπάς, γῆς ἀναδασμούς the age-old slogans of the revolutionary demagogue.[19]

It would not in the ordinary way be wise to build much social history on stock rhetorical clichés such as these - designed (of course) to discredit Procopius rather than draw attention to economic grievances. But there are other reasons for believing that economic factors did play an important part in the initial success of Procopius' bid for power.

Another speech of Themistius, his Pentaetericus (Or. viii, commemorating Valens' quinquennalia in March 368), draws a grim picture of the economic distress of the eastern provinces between Constantine and the accession of Valens (364). Throughout that period, he remarks, taxes went up steadily, eventually doubling after forty years.[20] Then came Valens, who after holding them at the same level for three years running, finally reduced them by half in the fourth. Forty years back from 364 takes us to 324, the very year Constantine defeated his last rival Licinius and set up his permanent establishment in the East, centred on Constantinople.

So Themistius is saying much the same as Anonymus. It was with Constantine that the increasing demands of public expenditure caused this calamitous rise in taxation and impoverishment of the lower classes.

At least half of Themistius' picture we may accept readily enough: the steady tax-rise since Constantine, and Valens' reluctance to increase the burden. Themistius' own qualification 'if the calculations go according to plan' (εἰ κατὰ γνώμην ἐκβαίνοι τὰ μέτρα τῆς ἐπανόδου)[21] reveals the projected 'halving' as perhaps no more than a pious hope, but for the rest there is the confirmation of the obituary notice in Ammianus (no admirer of Valens): tributorum onera studio quodam molliens singulari, nulla vectigalium admittens augmenta.[22]

But how soon did the privincials begin to feel the benefit - if benefit there was? According to Ammianus, it was not till he discovered how unpopular Valens had made himself by his rapacity (cupiditate aliena rapiendi succensum,[23] that Procopius decided to make his bid for the purple. In addition, Ammianus continues, Valens' agent, his father-in-law Petronius,[24] had been causing great distress by calling in arrears of tax from as far back as the days of Aurelian (debita iam inde a temporibus principis Aureliani,[25]

So even without a tax rise, if Valens was insisting on the full payment of arrears he must have made the immediate situation even worse. Those who were not up to date in their payments must have been faced with a tax demand which, by definition (if they had fallen into arrears in the first place), they would have been unable to meet. At such a moment 'abolition of debts' would have been a very seductive (and, for Procopius, very sensible) rallying cry.

Themistius' praise of Valens is not perhaps so straightforward as it seems. As G. Dagron has well put it[26] 'Thêmistios félicite de préférence les empereurs pour les qualités qui leur manquent manifestement le plus, ce qui est une manière de leur en recommander l'usage'. For example, most of Or. vii on the followers of Procopius praises Valens most fulsomely for his statesmanlike and conciliatory treatment of the rebels. Our other sources record a particularly savage and relentless persecution.[27] Themistius admits to writing several months after the final battle[28] and must have known the facts; he is doing his best to show Valens how, even now, he can best win men's respect and loyalty if he chooses.

So in the light of the very different story told by Ammianus and his own allusion to debts, it begins to look as if Themistius' emphasis on the success of the fiscal policy of Valens' first quinquennium in Or. viii may not mean quite what it says. Indeed, do not these suspiciously rosy remarks, bolstered up by impressive but misleading statistics, suggest rather that the beginning at least of Valens' reign saw a grave financial crisis?[29]

A crisis which (as so often) made the rich richer and the poor poorer. Themistius goes on to attack the rich for making even the barbarians welcome to their dependants[30] Anonymus also blames the rich for the dire straits of the poor, with harsh words for their houses crammed with gold (II.4). Perhaps, too, it is something more than a casual rhetorical exemplum when Themistius compares Procopius' methods of raising troops at some length to those employed by Spartacus and Crixus.[31] It is also interesting to note that, like Anonymus again, Themistius was tactful enough to evoke the then two year old Valentinianus Galates, remarking on the kingly qualities he was already revealing.[32]

Procopius and Marcellus fit Anonymus' <u>tyranni</u> down to the last detail. No other set of <u>tyranni</u> fits at all. Over and above this the general economic background to Themistius' <u>orationes</u> vii and viii (366/7 and 368) corresponds exactly to the situation described by Anonymus.[33]

Granted (as we surely must) that Anonymus wrote in the East and was addressing Valens, there is more than one pointer to a date before 370, when Valens' son died. The fortification of the Danube frontier seems to have begun in about 369;[34] it was being described with admiration by Themistius in 370.[35] It was in 370 that Valens finally invaded Persia in response to years of provocation.[36] One gets the impression from Anonymus that hostilities with Persia, though imminent, had not yet begun. Less precisely, the use made of Procopius' rebellion to illustrate the danger of current fiscal policies suggests that it was still relatively fresh in the writer's mind.

If we can narrow the limits to (say) 367-9, then there is perhaps a passage which might be pressed into a more precise reference. At VI.2 Anonymus refers to some of the barbarians who harass every frontier as occupying 'commanding mountain positions'. Now which barbarians could be said to have commanding mountain positions to retire into? I can think of only one such enemy: the Isaurians. In theory 'pacified' by Pompey, they were not perhaps technically barbarians, but in practice they constiuted an enemy enclave in Roman territory, launching periodic invasions of the surrounding provinces. One particularly wide ranging such incursion, 'inflicting great losses on Pamphylia and the Cilicians' and wiping out a vicar of Asia,[37] took place in 368, the first since 353 and the last before 376.[38] An army was dispatched, and 'after killing some of them drove the rest to the rocky retreats in the mountains where they live'.[39] If he had been writing in 368/9 it would be natural enough for Anonymus to have thought of the Isaurians when listing the natural advantages enjoyed by Rome's enemies.

If Anonymus had been trying to sell his inventions to Valentinian, he would surely have advertised the range of his <u>Ballista fulminalis</u> as the width of the Rhine rather than the Danube. It was the Rhine frontier that exercised Valentinian for most of his reign. Valens, however, spent the whole of the period from Spring 367 to Spring 370 on the Danube, making regular expeditions from his camp at Marcianopolis[40] against the Goths on the other side. There is of course no need to suppose that Anonymus was at Marcianopolis himself, though a personal approach would not have been a bad idea for one so anxious to bring his ideas to the emperor's attention - especially at a time when he was much occupied with military reform.[41] Indeed, it has not (I think) been noticed that, early in 370, Valens is reported by Themistius to have been equipping his new Danube forts with 'weapons and missiles *and engines*, all thoroughly tested' (ὅπλα καὶ βέλη καὶ μηχανήματα, ἅπαντα εἰς τὸ ἔσχατον ἐξητασμένα).[42] It would be nice to think (nor is it altogether impossible) that one or two of Anonymus' inventions - perhaps a *Ballista fulminalis* - were among them.

Anonymus' Latin, as has often been remarked, is very peculiar. Indeed, there has often been a feeling that his native tongue must have been Greek. That can be neither proved nor disproved;[43] it should be observed, however, that Latin would certainly have been the best language for a pamphlet aimed at Valens, for he did not know Greek. The nearest parallel is the *Breviarium* of Festus, a summary of Roman history (with special reference to the eastern provinces) addressed to Valens (presumably at Marcianopolis) in 369/70 by his *magister memoriae*.[44] The work closes with an evocation of the recent Gothic victory and a hope for success in Persia ('*ut ad hanc ingentem de Gothis* [sc. *victoriam*] *etiam Babyloniae palma pacis accedat*').

After concluding a peace with the Goths, Valens moved in 370 to Syria, where, on and off, he remained for several years, directing the Persian war. It looks as if Anonymus was writing *ca.* 368/9, when war with Persia was threatening and the Gothic war not yet over; when the building of the frontier forts had not yet really got under way; only a couple of years after the suppression of Procopius and Marcellus in 366, very soon after the Isaurian invasion of 368, and when the unfortunate Valentinianus Galates still had a year or so to live.

Department of Classics,
Colombia University,
NEW YORK N.Y.10027.

## NOTES AND REFERENCES

1. Thompson, 2, and works there cited.

2. S. Mazzarino *Aspetti sociali del quarto secolo* Rome (1951), 72-109- and indeed much of the rest of the book.

3. Amm. xxvi.5.13.

4. *ibid.* xxviii.3.4ff.

5. See the works cited by Mazzarino, p.86.

6. See pp. 60-1. But in the light of the parallel with Them: *Or*. vii adduced below I should be reluctant to correct Anonymus' error by writing 'Constantii' (the easiest of changes: cf. the first chapter of Ammianus Marcellinus, where 'Constantii' must be read for the 'Constantini' of the MSS - and often elsewhere).

7. Still less for his inference, based on the fact that Anonymus always uses *principes* to designate his plurality of emperors in the preface and *imperator* only for his main addressee, that only one of the *principes* was an *imperator*, that is to say Augustus (which would suit Constantius and his Caesars, but not the co-Augusti Valentinian and Valens). There is no basis whatever for this distinction, either in linguistic usage or constitutional practice.

8. Cf. the *Gesta Senatus Romani* in Mommen's edition of the *C Th*, i, p.1.

9. Gregorius and Hermogenianus, both under Diocletian; in general see J. Gaudemet, *La formation du droit séculier et du droit de l'église au IV$^{e}$ et V$^{e}$ siècles* (1937), and A. H. M. Jones, *Later Roman Empire* 472f.

10. *op.cit.* (n. 2), 385, n.66.

11. See now A. Demandt's useful discussion in *Hermes* 1972, 97f.

12. Amm. xxix.5.20; Zos. iv.16.2.

13. *PLRE* s. v. Galates.

13. The basic source is Amm. xxviii.6, with the discussions (quoting the earlier literature) of B. H. Warmington, *Byz. Zeitschr.* 49 (1956) 55f., and A. Demandt, *Byzantion* 38 (1968), 343f., *Historia* 18 (1969), 598f., and *Hermes* 100 (1972), 94f.

15. A. Piganiol, *L'Empire chrétien*$^{2}$ (1972), 172-4. I. Hahn, *Acta Antiqua* 6 (1958), 199-210 (in Russian: German abstract p.211) presents a rather simple insecurely based Marxist analysis of the rebellion.

16. Amm. xxvi.7.1.

17. *ibid* xxvi.7.7.

18. *ibid*. xxvi.10.3.

19. Them. *Or*. vii 91c = i.138.12 Downey.

20. *ibid*. viii.113a-c = i.170-1 Downey.

21. *ibid*. 113c.

22. Amm. xxxi.14.2.

23. *ibid*. xxvi.6.6.

24. *PLRE* Petronius no. 3.

25. Amm. xxvi.6.7; cf. *ibid*. 17 *Sepulta iam dudum negotia et redivivas nebulas* [trifles] *debitorum in diversos ordines excitabat*.

26. G. Dagron, 'L'Empire romain d'orient au IV$^{e}$ siècle et les traditions de l'hellénisme: le têmoignage de Thémistios', *Travaux et Mémoires* 3 (1968), 84, n.1.

27. Amm. xxvi.10.6f.; Zos.iv.8.5.; Soc.*EH* iv.5; Soz.*EH* vi.8; the only exception is Lib. *Or*. i.171, cf. xix.15f. and xx.25f., with A. F. Norman, *Libanius' Autobiography* (1965), 199-200. Themistius, like Libanius, presumably had a few particular examples of Valens' clemency in mind.

28. Them. *Or*. vii.84b = i.128 Downey.

29. Strongly borne out by Themistius' attempt in the main body of the speech to deter Valens, largely for fiscal reasons, from prolonging his offensive against the Goths (see the excellent analysis in Dagron, *op.cit*. (n.26). (101ff.)

30. Them. *Or*. viii.115a.

31. Them. *Or*. vii.86c. Cf. R. Browning, *CR* 1953, 107. It is of course an altogether different matter to claim, with I. Hahn (*Acta Antiqua* 1958, 199f.), that Themistius saw Procopius' uprising as a 'pendant' to that of Spartacus, with a 'social-revolutionary programme'.

32. Them. *Or*. viii.120a.

33. Though of course Anonymus takes for granted the aggressive policy towards barbarians that, even before Adrianople, Themistius consistently opposed, cf. Dagron. *op.cit*. (n.26), 84.

34. H. von Petrikovits, 'Fortifications in the North-western Roman Empire from the Third to the Fifth Centuries A.D.', *JRS* 61 (1971), 186ff., quoting the archaeological evidence - but mistakenly attributing the Danube as well as the Rhine fortifications to Valentinian.

35. Them. *Or*. x.136a-138d.

36. A. Piganiol, *L'Empire chrétien*$^{2}$ (Paris 1947), 176.

37. Amm. xxvii.9.6.

38. See J. Rougé, 'L'Histoire Auguste et l'Isaurie au IV$^{e}$ siècle, *REA* 68 (1966), 282f., and R. Syme, *Ammianus and the Historia Augusta* (1968), 43ff. - without necessarily accepting all their conclusions.

39. Amm. xxvii. 9. 7.

40. O. Seeck, <u>Regesten der Kaiser and Päpste</u> (Stuttgart, 1919), 231-9.

41. See the enthusiastic account of his reforms on the Danube frontier in Themistius, <u>Or</u>. x.135d f., and (more generally) Jones, <u>Later Roman Empire</u> 148-9.

42. Them. <u>Or</u>. x.136a.

43. See Robert Ireland, III 4: The style: a general summary.

44. See J. W. Eadie's edition London, (1967), 2-3, with my remarks in <u>CR</u> New Series 19 (1969) 305-6.

# THE ILLUSTRATIONS OF THE ANONYMUS, DE REBUS BELLICIS

by J. J. G. Alexander

The illustrations of the Anonymus, de Rebus Bellicis, are an essential part of it and, as Professor Thompson says, could not have been created with the aid of the text alone.[1] They are likely, therefore, to have accompanied it from the time of its original composition in the later fourth century. The treatise is transmitted to us, however, with a collection of texts which was probably put together in the ninth century.[2] Several of the other texts in the collection, especially that known as the Notitia Dignitatum utriusque imperii were also illustrated.[3] So far as is known at present all these illustrations were transmitted via a single Carolingian intermediary of the late ninth or early tenth century. This was a manuscript preserved in the cathedral library at Speyer (Codex Spirensis) of which a single, unillustrated, bifolium alone remains.[4]

The earliest copy of the Spirensis of which we know is also a fragment, the manuscript written in 1427 by Antonius Angeli de Aquila now in Cambridge.[5] Its five surviving leaves have pictures but are all from the Notitia Dignitatum. The earliest complete copies of the Spirensis including the illustrations of the de Rebus Bellicis are the manuscripts in Oxford (often referred to as 'O', but here, following Maier and Thompson, as 'C')[6] and Paris ('P').[7] 'C' was made for Pietro Donato, bishop of Padua, in 1436 when he was presiding for Eugenius IV over the Council of Basel. The colophon also states that he borrowed the Spirensis for it to be copied. The illumination has been convincingly attributed by Professor Sheila Edmunds to a Frenchman, Perronet Lamy, who was employed by Amadeus VIII, Duke of Savoy.[8] Perhaps Amadeus sent him to work for Donato with an ulterior motive, for in 1440 the Council elected Amadeus Pope in opposition to Eugenius, and he took the name Felix V. At any rate Lamy must have stayed in Basel a little time for he also executed the frontispiece of a Lectionary for Donato[9] and the illustrations of a Terence.[10] Moreover he also painted the miniatures of 'P' on which he must have worked at almost the same time. The Milanese humanist Pier Candido Decembrio made notes on this manuscript in his commonplace book in 1437 and it may have been made for him.[11] It is not so grand as Donato's copy to which Lamy added a frontispiece on two pages, which in 'P' is condensed into one, and a miniature of Constantinople not included in 'P', as well as making other alterations.[12]

Donato had been a pupil of Gasparinus Barzizzius who was professor in Padua in 1407, and he was a friend and correspondent of the Venetian humanist Francesco Barbaro. Guarino dedicated a Latin translation of Lucian to him. He had a fine library which is now dispersed but of which we have a catalogue.[13] Vespasiano da Bisticci wrote a life of him in which he accuses him of worldliness. Amongst his surviving manuscripts is a collection of classical material now in Berlin.[14] This includes inscriptions discovered by Cyriaco of Ancona[15] and Cyriaco added an excerpt from Gregory Nazianzus, _De septem mundi spectaculis_, in his own hand at the end of 'C'. Donato thus forms an important link in the tradition of Paduan humanism which begins with Petrarch's friendship with the Carrara overlords[16] and culminates in the second half of the fifteenth century with the distinguished circle to which, amongst others, the antiquarians Giovanni Marcanova and Fra Giocondo, the painter Andrea Mantegna, and the calligraphers Felice Feliciano and Bartolomeo Sanvito, belonged.[17] This tradition was throughout notable for an interest in the visual material of Antiquity.

Of the later copies of the _Spirensis_ by far the most important is that which belonged to the Count Palatine, Ottheinrich (1502-59), now in Munich.[18] This has two sets of illustrations, the first set belonging with the original manuscript made in 1542, the second set added at the end. A preface explains that Ottheinrich found the first set unsatisfactory due to their contemporary German Renaissance style and we know that the second set are in fact tracings of the _Spirensis._ Though the illustrations of 'M*' thus have great authority, we know that permission for the tracings was at first refused by the Chapter on the grounds that the _Spirensis_ was in a fragile condition.[19] This may have been an excuse, of course, to avoid loaning the manuscript, but it cannot be assumed in all cases that the 'M*' pictures are more accurate than those in 'C' and 'P', for the _Spirensis_ may have been in better condition in the fifteenth century. There are a number of other illustrated copies, but those mentioned are the most important witnesses.

On the whole the Anonymus illustrations are similar in 'M*' and 'O' and this is a tribute to the accuracy with which Lamy performed his task. Dr. Reece has shown (p. 20) that illustrations of the Anonymus' proposals on the reform of the coinage are altered in 'C' by Lamy who inserts representations of early imperial coins which cannot have been in the original and which are not shown in 'P' or 'M*' (Fig. 3). On the mast of the _Liburna_ where 'M*' shows a cross with a classical-looking wreath, Lamy omits the wreath in 'C' and both wreath and cross in 'P'. On the other hand Lamy paints an eye on the prow which is not shown in 'M*' but which might perhaps be authentic (Part 2, Pls XX a, b, XXI a, b). On the wings of the _Currodrepanus clipeatus_ small tassels are represented in 'C' but not in 'M*'. It is interesting, however, that in 'M' they are shown, which might suggest that these illustrations were copied not from the _Spirensis_ but from some intermediary influenced by 'C'. (_Ibid._ Pls. XIV a,b, XV a, b).

If the Renaissance copies give us a fairly accurate idea of the illustrations of the _Spirensis_, we need to ask what earlier stages there were in their transmission. It seems likely that the _Spirensis_ copied an earlier Carolingian model rather than a late antique manuscript direct. Since the early Carolingian copies of classical texts are remarkable for their ability to reproduce the illusionistic style of their models, the artist of the _Spirensis_ was no doubt responsible for the unclassical stylization of the pictures as transmitted in 'M*'. The late antique model, since it probably contained the _Notitia_ as well as the Anonymus, is likely to have been of the early fifth century perhaps produced in the same centre as the Vatican Virgil, that is Rome.[21] Good comparisons can be made between the Virgil illustrations and the pictures in the *Notitia*. The early fifth-century manuscript must have been a very exceptional product, a *de luxe* copy made for some great personage, perhaps the Emperor himself.

If we compare the illustration of the Anonymus and those of the *Notitia*, though in both we can detect a similar formal language, there are certain significant differences. Firstly, whereas most of the *Notitia* pictures are full-page miniatures, the Anonymus illustrations are rectangular pictures inserted into the text. In a number of pictures objects overlap the frame as if the original composition was not intended to be restricted in this way. Examples are the spear in the *Thoracomachus* picture (Part 2, Pls. XVI a, b, XVII a,b), the mast of the *Liburna* (*ibid*. Pls. XXa, b, XXI a,b), and the scythe on the chariot of the *Currodrepanus clipeatus* (*ibid*. Pls. XIV a,b, XV a,b). Secondly, most of the *Notitia* pictures are symbolic. They represent not the great officers of the Empire in person but their insignia, the trappings of their office. These are shown diagrammatically in a way typical of Roman art, scattered in a haphazard manner and seen in a birds-eye view rather like the famous *asaroton* mosaic pavements. Some of the Anonymus pictures are also diagrammatic birds-eye views, for example the *Ascogefyrus* (Part 2, Pl. XVIII a, b, XIX a, b). But the majority are landscape views from a normal view-point showing the machines in use by the soldiers. For example in the miniature of the *Thoracomachus* (*ibid*. Pl. XVI a, b, XVII a, b) a soldier stands wearing some of the equipment whilst other pieces are hung on a convenient tree. A realistic context for the objects is thus provided.

If the distinctions drawn are valid, then they may not be without significance for determining the nature of the original illustrations of the archetype. The fourth century was the time when, as Professor Weitzmann has shown, the nature of book illumination started to change to take account of the replacement of the roll by the codex as the normal type of book.[22] Unframed miniatures set in the columns of text are the normal form of illustration in the papyrus roll. Such illustrations continued to be used later in codices, of course, for example in such Eastern illuminated manuscripts as the Vienna Genesis of the sixth century. The original Anonymus illustrations on this evidence, whether the archetype was a roll or a codex, could have been 'roll-type' illustrations set in the texts and, assuming the overlapping of the frames is not due to the Carolingian copyist, lacking frames. If so they might have been rather more likely to originate in an Hellenistic or eastern influenced milieu than a western one. The Anonymus text with its illustrations would then have been incorporated into the early fifth-century codex whose major component was the Notitia mainly illustrated by 'codex-type' full-page pictures. Frames may have been added to the Anonymus pictures at this stage in order to make them more consistent.

History of Art Department,
The University,
Manchester. M13 9PL

## NOTES AND REFERENCES

1. Thompson, espec. pp. 15-17.

2. L. Bieler, 'The text tradition of Dicuil's Liber *de mensura orbis terrae*', *Proceedings of the Royal Irish Academy*, 64, section C no. 1, (1965), 1-32.

3. The edition of O. Seeck, *Notitia Dignitatum* (Berlin, 1876), reproduces the illustrations of the Munich MS. in outline. For the textual problems of the *Notitia* and the extant MSS. see I. Maier, 'The Giessen, Parma and Piacenza Codices of the "Notitia Dignitatum" with some related texts', *Latomus*, 27 (1968), 96-141; *id.* 'The Barberinus and Munich codices of the *Notitia Dignitatum omnium*', *ibid.* 28 (1969), 960-1035. For the illustrations see J.J.G. Alexander, 'The illustrated manuscripts of the *Notitia Dignitatum*', in R. Goodburn and P. Bartholomew (eds.) *Aspects of the Notitia Dignitatum*, B.A.R. Supplementary Series 15 (Oxford, 1976), 11-25.

4. Schloss Harburg, Fürstlich Oettingen-Wallerstein'sche Bibliothek, I, 2, 20, 37. For references and an illustration see Alexander, *op. cit.* (n. 3), Pl. I.

5. Fitzwilliam Museum, MS. 86-1972. It was bequeathed by Professor Francis Wormald. It was published by H. Omont, 'Le plus ancien manuscrit de la Notitia Dignitatum', *Mémoires de la société nationale des antiquaires de France*, ser. VI, 1 (1890), 225-44. It had belonged to Sir Thomas Phillipps. MS. 16397, and Guglielmo Libri. For its earlier history see Alexander, *op. cit.* (n. 3). The Anonymus illustrations were already known to Konrad Kyeser who wrote his treatise *Belli Fortis* 1402-5. Thompson, 18.

6. Bodleian, MS. Canon. Misc. 378. For the illustrations see Part 2 below. For bibliography see O. Pächt, J.J.G. Alexander, *Illuminated manuscripts in the Bodleian Library, 2, Italian School* (Oxford, 1970), no. 599.

7. Bibliothèque nationale, latin 9661. H. Omont, *Notitia Dignitatum Imperii Romani. Reproduction réduite des 105 miniatures* (Paris, 1911). For the illustrations see also Part 2 below.

8. S. Edmunds, 'The Missals of Felix V and early Savoyard illumination', *Art Bulletin*, 46 (1964), 138-40.

9. New York, Morgan Library, M. 180.

10. Vatican, Ottob. Lat. 1368. This was written at Basel in 1436 by the same unidentified scribe as 'C' and 'P'. The original owner is unknown. The arms are unfortunately defaced.

11. R. Sabbadini, 'Spogli Ambrosiani Latini. Di un nuovo codice della "Notitia Dignitatum"', Studi italiani di filologia classica, 11 (1903), 257-63.

12. See Alexander op. cit. (n. 3), 15 and Pl. X.

13. P. Sambin, 'La biblioteca di Pietro Donato (1380-1447)', Bolletino del Museo Civico di Padova, 48 (1959), 60.

14. Berlin, Deutsche Staatsbibliothek, Hamilton 254. H. Boese, Die lateinische MSS. der Sammlung Hamilton zu Berlin (Berlin, 1966).

15. B. Ashmole, 'Cyriac of Ancona', Proceedings of the British Academy, 45 (1959), 25-41. W. Bodnar, Cyriacus of Ancona and Athens (Collection Latomus 43 (1960).)

16. F. Saxl, 'Petrarch in Venice' and 'Jacopo Bellini and Mantegna as Antiquarians' reprinted in A Heritage of Images. A Selection of Lectures by Fritz Saxl, ed. E. Gombrich (Harmondsworth, 1970).

17. C. Mitchell, 'Felice Feliciano Antiquarius', Proceedings of the British Academy, 47 (1961); J. Wardrop, The Script of Humanism (Oxford, 1963).

18. Munich, Bayerische Staatsbibliothek, Clm. 10291. For the two sets of illustrations in 'M' see part 2 below.

19. Maier, op. cit. (n. 3, 1969), 995.

20. See p. 65.

21. Vat. Lat. 3225. J. de Wit, Die Miniaturen des Virgilius Vaticanus (Amsterdam, 1959).

22. K. Weitzmann, 'Book illustration of the fourth century: Tradition and Innovation', in Studies in Classical and Byzantine Manuscript Illumination, ed. H. L. Kessler (Chicago, 1971), 96ff.

# LATE ANTIQUE BOOK ILLUSTRATION AND THE GALLIC PREFECTURE

by Martin Henig

This paper has a double origin. I was originally asked to say something about the illuminations in the Anonymus, de Rebus Bellicis. However, my thoughts were soon diverted by the great silver dish known as the Corbridge Lanx, into considering how people approached pictures in Late Antiquity. The spirit of the Lanx, "a topical ad hoc allusion to a specific event, namely to the sacrifice offered to Apollo on Delos by Julian the Apostate in 363"[1] seemed to me especially reminiscent of the Virgilian illuminations in Codex Vaticanus Latinus 3867[2]. Both the works, particularly the lanx, are close in date to the original versions of the illustrations of the Anonymus, and it is against the Canons of Late Antique Art they represent, that the pictures of the de Rebus Bellicis should be set. A study of the origins of this style has particular relevance for our author since technical treatises would appear to have been among the earliest works to be illustrated.[3] Such a study forms the first part of my paper. Yet the Lanx and the Vatican Vergil were products of an imperial or metropolitan milieu; and if one thing is agreed about the Anonymus, it is that he was a provincial. How prevalent then, was the practice of book illumination in provincial studios? In attempting to answer this question in the second part, I examine one area in particular, the Gallic prefecture, whose role in this sphere has previously been at best underestimated and at worst ignored.

## Late Antique book illumination and its development

The art of book illumination had humble beginnings as simple diagrams which were employed to clarify the meaning of scientific and mathematical treatises.[4] These works were written on papyrus scrolls with the text arranged in columns and the figures occupying convenient breaks. Although the earliest surviving example of this practice is hellenistic in date, it is thought to have originated in classical times.

Good examples of such diagrams, embellished sufficiently to deserve the name of art, were copied in the surveying treatises of the Corpus Agrimensorum, which includes representations of cities in classical style.[5] These may be compared with the town on a Pompeian fresco showing the fall of Icarus,[6] and with another in the fourth century Vergilius Vaticanus (fol.35$^{v}$) as well as with those shewn in the Notitia Dignitatum. Mediaeval copies of Dioscurides' de Materia Medica incorporate plant illustrations of such verisimilitude that there can hardly have been many intermediary copies since the first century original.[7] Finally, the siege engines and automata which appear in works under the general heading of Poliorcetica and Pneumatica[8] (Pl. I) from the second century B.C., are in the same tradition as the devices portrayed in the de Rebus Bellicis. They stand alone rather than as items in

PLATE I. Siege Engine. Vatican, Biblioteca Apostolica Vaticana, Vat. Gr. 1164, f. 99r

full-scale miniatures. Thus, in the Anonymus, the representations of the Tichodefrus and Clipeocentrus (Part 2 Pls. VII, VIII) of the lances, plumbata et tribulata and plumbata mamillata in their box (ibid., Pl. IX) and of the skin-bridge known as the ascogefrus (ibid., Pls. XVIII, XIX) are very close indeed to ancient prototypes. Elsewhere, with one exception, the human figures appear as mediaeval intrusions. This one exception, however, proves the rule for the machine is an item of underwear - the Thoracomachus (ibid., Pls. XVI, XVII) and the Roman Soldier is an essential part of the explanation of its use.

Current orthodoxy maintains, I believe, that fullscale pictures scarcely occurred in manuscripts until long after the codex was invented in the early centuries of our era. This view does not seem to me to be consistent with the evidence of wall painting in general and of the Odysseus frieze from Rome in particular.[9] That frieze, executed in the first century B.C. and based on a second century original, is remarkable for the way in which the figures are identified - minutely - in Greek which would have been too distant to read when the frieze was high on a wall. Other wall paintings depicting mythological scenes are repeated in identical form at several sites and suggest that the artist had access either to pattern-books or, more probably, to illustrated texts.[10] It is of some interest that Martial writes of a scroll which was prefaced by a portrait of Vergil (XIV, No. 186) and presumably this would have been similar to those that still survive in the Vergilius Romanus.

In any case, there can be little doubt that illustrated scrolls and codices existed in considerable numbers by the third century A.D., for there is a direct iconographical link between the Monnos mosaic at Trier and the classically derived twelfth century astronomical codex in Madrid[11] whilst the depiction of Andromeda chained to a rock in a ninth century manuscript at Leiden is based on a very well-known prototype.[12]

Actual manuscripts survive from the fourth or fifth centuries. Of primary importance are the Vergilius Vaticanus (Cod. Vat. Lat. 3225) which is highly eclectic, but frequently very classical in style, and the vigorous and rather strange Vergilius Romanus (Cod. Vat. Lat. 3867) which with Grabar[13] I believe to be not only a finer work of art but every bit as early in date (Pls. II a,b, VIII b). Its style seems to be neither Oriental nor early Byzantine and in the second part of this paper I will attempt to place it in the latin-speaking culture of Gaul and Britain. These documents may be related to other contemporary works of art as well as to books slightly later in date (e.g. the Ambrosian Iliad) and also to others which are renaissance or early modern copies of Late Antique books (e.g. Notitia Dignitatum, Pl. VII, Filocalus' Calendar of 354, Pls. IV, V, VI).

Thus, to take one theme, many of the illustrations in the calendar include an assortment of objects scattered in the field; for example, the City of Trier[14] is personified as a goddess of Minerva type with her right hand on the head of a barbarian. Her warrior aspect is further suggested by two shields and a quiver full of arrows while her wealth is symbolised by several drinking vessels. The same pictorial shorthand is found in the Vergilius Romanus[15] and in the Notitia Dignitatum.[16] It stems from the "Asaroton" tradition which can be traced back in time, via the mosaics of S. Constanza in Rome and the frescoes of the Dura Europos Synagogue to the work of Sosos.[17] The only difference between later examples of the type and their hellenistic models is their apparent flatness and avoidance of perspective.

Again, both new and archaising elements can be found in the consular diptychs, which are in essence illuminated pages fossilised in ivory, and in silver plate. The pictures of Gallus proffering a victory to Constantius II who is in the act of distributing largesse are the most formal and hieratic in the *Calendar*.[18] They find their exact parallels in the diptychs of Probus[19] and Probianus,[20] and in the Missorium of Theodosius[21] where the dignity of the Roman magistrate and even more of the sacred Emperors is maintained in the same, elevated, late antique manner. As a contrast the famous Nichomachus-Symmachus diptych[22] which depicts two priestesses and a similar pagan diptych[23] that shows Asklepios on one leaf and Hygeia on the other, or the Corbridge Lanx[24] are as classicising as the representations of the months in the *Calendar*[25] and belong to a much older tradition altogether, a tradition epitomised by the mythological reliefs on the Ara Pacis.[26]

## The Gallic Prefecture

Is it possible that drawing and illumination was carried on in North Western Europe, even perhaps in Britain? If so, under what conditions? Such questions require no excuse because only a few centuries later - admittedly in a rather different environment than that which appertained in Roman Britain - Celts and Anglo-Saxons were producing some of the greatest works in the history of mediaeval illumination. The products of *scriptoria* at Iona or Lindisfarne were early and influential and far finer than contemporary products emanating from the old lands of the Roman Empire. Books such as the *Codex Amiatinus*, the *Lindisfarne Gospels* and the *Book of Durrow*[27] were produced for monastic patrons whose place in their rich and settled communities was similar to that of the late Roman (noble) *patronus*. Indeed, as Richard Reece has so rightly pointed out to me, there may be a question of survival or continuity to be investigated between the Late Antique period and the 'Northumbrian Renaissance'. In the Britain of the fourth century the thriving culture of the Great Villas offered propitious ground for artistic endeavour of all kinds.

It is to Rutilius Namatianus, an inhabitant of Gaul writing in the early fifth century A.D., that we owe the finest and most moving tribute to Rome 'mother of men and mother of gods'.

> 'Exaudi, regina tui pulcherrima mundi,
>     Inter sidereos Roma recepta polos;
> Exaudi, genetrix hominum genetrixque deorum;
>     Non procul a caelo per tua templa sumus.
> Te canimus semperque, sinent dum fata, canemus
>     Sospes nemo potest immemor esse tui .....
>
> * * * * *
>
> Fecisti patriam diversis gentibus unam.
>     Profuit iniustis, te dominante, capi;
> Dumque offers victis proprii consortia iuris,
>     Urbem fecisti, quod prius orbis erat.'

*De Reditu Suo* 1, 47ff.

As Rutilius suggests the lands of the Empire shared in a common culture. Even the north-western provinces, once the home of rude and barbarous peoples had become an important part of the Graeco-Roman oikoumene; a process whose beginnings are recorded by Tacitus (Agricola 21). For a time in the third and fourth centuries Trier was an imperial captial and, inevitably, a centre where arts and letters were patronised. The poems of Ausonius on the one hand, and the ceiling paintings from the Constantian palace[28] on the other attest a sophistication equal to that of Antioch or Alexandria or Milan.

High culture was not confined to one city,[29] nor did it stop at the Channel. Great villas such as Woodchester, their floors embellished with mosaics, dinner services of exquisite silver like that from Mildenhall, the incisive criticism of Augustine's great rival Pelaguis,[30] and the success of the local curial class in continuing the administration for more than a generation after links with the central Empire were severed at the beginning of the fifth century[31] together comprise strong evidence in favour of a rich and cohesive society. Indeed, if a rescript issued by Gratian and Valentinian in A.D. 376 and addressed to the Prefect of the Gauls means what it says (C.Th. XIII, 3, 11), provision was made for the support of Greek and Latin rhetoricians in the major cities of the praefecture, which assuredly must have included London.... so much for the primacy of Oxford and Cambridge in University education in this country, unless one accepts the fantastic claims of sixteenth and seventeenth century storytellers, who go back to Cantaber and the era of the Trojan War!

But it is to this war, one of the most memorable in the history of mankind both for its sombre and heroic events and its unexpected yet glorious aftermath, that I must now turn, or rather to evidence for the popularity of Virgil in Britain. Evidently Carausius intended his coins with the legend EXPECTATE VENI to evoke the Virgilian, 'Quibus Hector ab oris exspectate venis' (Aen. iii, 282f.), while an inscription above the Europa mosaic at Lullingstone is a scholarly joke alluding to an episode in the first book of the Aeneid (I, 88ff.):

> "invida si ta(uri) vidisset Iuno natatus,
> iustius Aeolias isset adusque domos."

The great storm which Juno raised against Aeneas is graphically shown in one of the miniatures of the Vergilius Romanus (fol.77$^{r}$). Of course it is rather simplistic to suggest that the epigram on the floor of a triclinium was intended to evoke discussion at every dinner-party (unless the owner was accustomed to boring his guests), yet it was surely designed to give a cultivated tone to the proceedings.

Scenes from the Aeneid were shown on a fresco painted on the wall of a room in the villa near Otford, Kent[32] as the inscription painted in rustic capitals, 'Bina manu l(ato crispans hastilia ferro)', testifies (cf. Aen. i, 313; xii, 165). There is an overwhelming probability that the scene accompanying the legend was copied from an illuminated Virgil, for Virgilian scenes in monumental painting are rare even in Italy, and I think it very unlikely that a workshop copy-book, even if it existed, would have contained texts as well as pictures. As, in fact, two illustrated ancient manuscripts of Virgil survive, any other explanation seems rather pointless.

The most valuable piece of evidence relating to Virgil from Britain is the mosaic from Low Ham mosaic, Somerset.[33] The scenes illustrate books I and IV of the Epic. The Trojan fleet, blown off-course in the tempest we have described above, comes to Carthage. Achates is then presented, with a spear held over his left shoulder and a diadem in his right hand. He appears in identical form (but without the diadem) in the *Vergilius Vaticanus* (f. $16^{r}$). Beynond, Ascanius and Aeneas stand before Dido; between the Trojans and the Queen is the rather sinister figure of Venus. Here we may note the iconographical similarity of Aeneas to Turnus, standing cross-legged and holding a spear in the *Vergilius Romanus* (Pls. II b, III b). Then we come to the Royal Hunt; Ascanius, Aeneas and Dido on horseback, which is, however, not related to any scene in the *surviving* manuscripts. The composition of the scene in the "darksome cave" here rendered as Dido and Aeneas embracing between two trees, although far removed from the corresponding scene in the *Vergilius Romanus* (f. $106^{r}$) is similar in composition to the rendering of Meliboeus and Tityrus between trees in f. $1^{r}$ of the same manuscript. Rosenthal's theory that mosaics like this influenced manuscript illumination is most unlikely because the breakdown of the narrative into single framed episodes, vital for a codex, is somewhat inconvenient in a mosaic; furthermore the transmission of artistic ideas demands a portable medium such as a book.[34]

For Rosenthal the *Vergilius Romanus* is an orientalising production which contrasts with the late classic forms of the Low Ham mosaic. Furthermore, it is a hundred and fifty years later in date. In contrast, I maintain that although there are few *iconographic* links between the two, there are important *stylistic* similarities. They come from the same *milieu.*

The attention paid to outline, the absence of perspective encouraging the development of rhythmic patterns, and the primitive vigour of the figures are features both have in common and which contrast with the paintings in the more correct and classical *Vergilius Vaticanus.* However, in each of these respects, the manuscript appears stranger and less classical than the mosaic.

If the *Romanus* is fourth century in date, it was produced in a Roman world dominated by the four cities shown in the Filocalus Calendar (Rome, Constantinople, Alexandria and Trier), so we should not be biased against ascribing even important court productions to the city on the Moselle. The use of Latin in the East was very restricted as the rarity of Latin papyri outside military circles testifies. That alone argues against the *Romanus'* Eastern provenance. Yet compare the portraits of Virgil (fol. $3^{v}$; etc.) with the surveyor in the Wolfenbüttel Codex of the Agrimensores (fol. $67^{v}$) and contrast the provinciality of the former with the high classicism of the latter, which is known to be of Italian workmanship.

Even more telling are the illustrations in the Calendar[35] which were originally drawn by a certain Furius Dionysius Filocalus (Pl. IV) and are clearly metropolitan in style. The goatherd who represents March and February clad in her chiton and himation (Pl. V) could have been adapted from 'Pompeian' frescos and the former is far removed from the more stylised countrymen in the Eclogue illustrations of the *Vergilius Romanus* (fol. $1^{r}$, $6^{r}$, $11^{r}$ (=Pl. II a) and $16^{v}$).

The Vergilius Romanus with its primitive and vigorous approach is certainly provincial and at least as Western as the little blond-haired German page-boys in the banquet (fol. $100^v$) who, incidentally, have their counterparts in some of the herms from Welschbillig,[36] although, of course, they could also be paralleled on works of art eminating from the Eastern Court. But, I believe, we can go further and ascribe the manuscript to Britain where Vergil was so widely read and loved. Certainly the figure-style and the linear treatment of draperies is consistent with all that we know of Romano-British art. The Virgil portraits (fols. $3^v$ (=Pl. VIII b), $9^r$, $14^r$ are highly reminiscent of a well-known relief sculpture from Murrell Hill, Cumberland[37] (Pl. IX) which shows a woman seated towards the front. "The main charm of the piece must always have resided in the ornamental black-and-white effect produced by the deeply gouged out lines of the trailing tunic." George Henderson has already demonstrated that the sculpture recalls St. John on fol. $209^v$ of the Lindisfarne Gospels,[38] but the parallel between the Gospels and Virgil would be just as striking and, of course, poses the question of manuscript survival as detailed above. The provincial treatment of dress in the Virgil portraits is also strikingly similar to that of the tunica and pallium worn by Christ on the Hinton St. Mary pavement (Pl. VIIIa), as John Peter Wild has pointed out to me.

The jet pendants made in Yorkshire around the time of the tetrarchy are in some ways even more significant because they required a miniaturist approach to the craft.[39] Before any engraver would attempt to cut such a hard material as jet, he is likely to have prepared very exact drawings of exactly what he intended to do. The reason the designs of the pendants are so like those of gold-glass emblemata is surely this common graphic source. It will be observed that the British pendants are very formalised and linear compared to contemporary cameos from elsewhere in the Empire.[40]

At a slightly earlier period we have the same problem with the types of Carausius' coinage. The die-cutting achieved a very high standard but many pieces exhibit a curious though (I hasten to add) aesthetically satisfying linearity.[41] Perhaps British drawing has something to do with it, even where a direct freestyle was employed in barbotine pottery[42] the final result has a strongly linear character.

Although the Romanus is the only Roman manuscript to which I would assign a possibly insular provenance, the original of the Notitia also appears to me as a product of the Gallic Praefecture. The Codices which contain it are dependent on Carolingian archtypes themselves descended from late Antique manuscripts. In style these probably had more in common with the provincialism of the Romanus that with works in 'metropolitan' style like the Filocalus Calendar. However, the illuminators are even further removed from the illusionism of such Eastern works as the Ambrosian Iliad. To take one example, the assembled deities shown in the Romanus ($234^v$ and $235^r$) are not unlike miniatures IX and XXIII of the Ambrosian Iliad from the point of view of composition. In each case the gods and goddesses are nimbate, shown frontally and wear or carry appropriate dress or attributes. Yet the Greek manuscript[43] is airy and impressionistic, recalling the painted shields from Dura Europos[44] while the Western one is solid and linear. The goddesses are reminiscent of images of Roma in the Notitia of the City of Rome[45] and

Campania in the Notitia Dignitatum[46] as well as of the female deities on the Brantingham mosaic.[47] Indeed, Roma, shown with a head-dress of 'spikes' in the Notitia, is paralleled by the Minerva in Romanus fol. 235r, whose plumed helmet has merged with her nimbus to create just such an effect. This is far removed from the careful depiction of Roma's helmet in the Calendar or on fresco painting from Rome.[48] We must bear in mind the possibility that Roma's head-dress is intended to suggest the ears of corn which are appropriate to Annona, although I do not think that this is likely.

Another example of differences arising between central and provincial studios can be seen in comparing Alexandria as represented in the Calendar with her equivalent in the Notitia. Alexandria (Pl. VI) is presented as a standing female figure who holds corn-ears and poppy-heads and is crowned with corn-ears. Behind her are two windows through which grain-ships are visible. The conception, highly regular and classical, was borrowed from personifications of Carthage and is the basis for the miniature with the insignia of the Proconsul Africae in the Notitia (Pl. VII). However, the plants which the goddes holds are blundered, as palms, and the ships now appear in a separate panel below, hence depriving the picture of any perspective.

A final note, apropos of the 'asaroton' effects to which I have referred above (p. 19) . The leaves of precious metal used as largesse on both the Vergilius Romanus and the Notitia bring to mind the votive tablets from Barkway, Hertfordshire,[49] and even more, the recently discovered 'christian' examples from Water Newton.[50] There was a considerable quantity of silver in the west at this time, and apart from actual plate we may cite the ingots found especially in the British Isles and perhaps used in the payment of troops.[51]

As I said at the beginning of this paper, my enquiry began with the Corbridge Lanx and with the de Rebus Bellicis and it is time to return to them. The Lanx was copied from a cartoon, presumably in some studio on the shores of the Mediterranean although it is still not easy to assign much Roman silver to specific schools or workshops and in view of the preponderance of silver from fourth century Britain, it would not be surprising if important officinae existed in Trier or even further West. Some of the items in the Mildenhall Treasure are said to be Gallic[52] and Sonia Hawkes has ascribed a silver bowl from Coleraine to Britain.[53] I feel that there is a strong and solid Western element in the Lanx, but this is not essential to our argument. The Lanx proves that certain visual effects found in manuscripts were known in Britain, and that the aesthetic consciousness required to appreciate high-quality drawing existed even in the north of the province. Pelagius and his contemporaries would have known of similar vessels in their youth just as they knew of books and learned how to write with elegance and panache.

The same 'asaroton'-like features are found on the Lanx as on late Roman illuminations and, furthermore, the figures are positioned in relation to trees in the same way as on such manuscripts (e.g. the Tityrus episode in the Romanus fol. 1r and appropriate pictures in the Vaticanus, fol. 36v, 69r), pagan diptychs and sarcophagi. Dishes such as the Corbridge Lanx belonged to members of the same landowning class of society which dined in triclina floored with Virgilian pavements and read the classics with ease. Such people must have commissioned illuminated codices even if they did not physically create them (and the name of the illustrator of the *Calendar* suggests an *honestior* at the very least). The Roman Empire offered many opportunities for artistic diffusion between East and West and if it is strange to think of a silver dish engraved with the sacred representations of the gods and goddesses worshipped on the most evocative of Greek islands, floating downstream on the muddy floodwaters of the Tyne, it is no more curious that the lands which later gave birth to the *Codex Amiatinus* and the *Lindisfarne Gospels* should have made an earlier contribution to the Civilised Life.

This brings us at last to the *de Rebus Bellicis*. It is difficult to say exactly where our Author wrote, although the references to the building of *burgi* brings Valens frontier policy to mind

This brings us at last to the *de Rebus Bellicis*. It is difficult to say exactly where our Author wrote, although the references to the Danube and the building of *burgi* brings Valens' frontier policy to mind (see pp. 6 and 70). However, for our purposes, it does not matter whether the tract actually originated in the East or in the West: technical drawing was certainly employed in both sections of the Empire; in the West this can be proved from the existence of the Cadestral Survey of Orange on marble[54] while a relief from Buzenol showing a complicated reaping machine called a *vallus* is close to the tradition of the Anonymus (p. 39, Pl. Xa, b). At an early date the Anonymus' treatise certainly existed in the West as the other works with which it was associated—all in Latin—imply. There is no reason why new developments in technology should necessarily leave their traces in art (although no one really supposes that major buildings were ever erected without the aid of detailed plans). The illustrations of the *de Rebus Bellicis* escaped oblivion in that these ephemeral sketches were copied and these copies came to be bound up with the magnificant Notitia manuscript. Thus what had been merely utilitarian sketches have come down to us as precious evidence for the art of drawing in the Late Roman Empire.

Institute of Archaeology
Beaumont Street
Oxford.

## POSTSCRIPT

Since writing the above Rosenthal's *The Illuminations of the Vergilius Romanus* has been reviewed by Professor J. M. C. Toynbee.[55] Her main conclusions are that the work is of late fifth century date and Western. She suggests that the Romanus was 'painted for a wealthy and cultivated villa-owner in Gaul'. Together with Rosenthal she accepts the relevance of the Low Ham mosaic in the development of book illumination.

## NOTES AND REFERENCES

The author would like to thank, among others, Mark Hassall, Julian Munby and Richard Reece for various comments and suggestions made when an earlier version of this paper was read to the Roman Department of the Institute of Archaeology, London.


1. J. M. C. Toynbee, Art in Britain under the Romans (Oxford, 1964), 306ff.
2. E. Rosenthal, The Illuminations of the Vergilius Romanus (Zürich, 1972).
3. H. Stern, Le Calendrier de 354 (Paris, 1953); For Late Antique Book illustration and its origins see especially R.B. Bandinelli, Hellenistic-Byzantine Miniatures of the Iliad (Olten, 1955); K. Weitzmann, the works cited in no. 4 and 12.
4. K. Weitzmann, Ancient Book Illuminations (Cambridge, Mass., 1959). 5ff.
5. O. A. W. Dilke, The Roman Land Surveyors: an Introduction to the 'Agrimensores' (Newton Abbot, 1971), 126ff.
6. J. M. C. Toynbee, The Art of the Romans (London, 1965), 118.
7. K. Weitzmann, op. cit. (n.4) 12.
8. ibid, 7-10.
9. J. M. C. Toynbee, op. cit. (n. 6), 113ff.
10. G. M. A. Richter, Ancient Italy (Ann Arbor, 1955), 78 ff.
11. R.B. Bandinelli, op. cit. (n. 3), 18.
12. K. Weitzmann, Studies in Classical and Byzantine Manuscript Illumination (London &Chicago, 1971), 97ff; G. Picard, Roman Painting (London, 1970), 64 ff.
13. A. Grabar, Byzantium from the death of Theodosius to the rise of Islam (London, 1966), 194.
14. H. Stern, op. cit. (n. 3), 142f.; E. M. Wightman, Roman Trier and the Treveri (London, 1970), 65 and Pl. X.
15. Cod. Vat. Lat. 3867 f74$^{v}$ quiver, 76$^{v}$ diadem and leaves (?palms) of precious metals given as gifts; see E. Rosenthal, op. cit. (n. 2), Pls.
16. e.g. Oxford Bodleian Library MS. Canon. Misc. 378 f. 141$^{r}$ (weapons), ibid. f. 142v, 144$^{r}$ leaves and other appurtenances to largesse, illustrated in R. Goodburn and P. Bartholomew, Aspects of the Notitia Dignitatum (BAR Supplementary Series 15, Oxford, 1976), 1, Pl. XXVIII and XXVI and XXVII.

17. W. Dorigo, *Late Roman Painting* (London, 1871), 95 and 212; H.P. L'Orange and P.J. Nordhagen *Mosaics* (London, 1966), 37.

18. H. Stern, *op.cit.* (n.3), 152-68, Pls. XIV ff.

19. R. Delbruck, *Die Consulardiptychen* (Berlin-Leipzig, 1929), No.1.

20. *ibid.*, no. 65.

21. *ibid.*, no. 62 - W.F. Volbach, *Early Christian Art* (London, 1961), 322, No. 53.

22. R. Delbruck, *op.cit.* (n.19), no. 59 = Volbach *op.cit.* (n.21), 328, No. 90ff.

23. R. Delbruck, *op.cit.* (n.19), No. 55.

24. J.M.C. Toynbee, *Art in Roman Britain* (London, 1962), 172, No. 108.

25. H. Stern, *op.cit.* (n.3), 232-298.

26. E. Simon, *Ara Pacis Augustae* (Tübingen, 1967), 22-30.

27. R.L.S. Bruce-Mitford, 'The Art of the Codex Amiatinus' *JBAA*[3] 32 (1969), 1-25; E.G. Miller, *The Lindisfarne Gospels* (London, 1923); F. Henry, *Irish Art in the Early Christian Period* (London, 1965), 159ff.

28. W. Reusch, *Frühchristliche Zeugnisse im Einzugsgebiet von Rhein und Mosel* (Trier, 1965), 240-6.

29. cf. T.J. Haarhoff, *Schools of Gaul* (Johannesburg, 1958).

30. For Pelagius cf. J.N.L. Myres *JRS* 50 (1960), 21-36.

31. Cf. J. Morris, *The Age of Arthur* (London, 1973), *passim*.

32. J.M.C. Toynbee, *op.cit.* (n. 1), 220.

33. *ibid.*, 241-6; E. Rosenthal, *op.cit.* (n. 2), 106-9.

34. *ibid.*, 109.

35. Vatican MS. Barb. Lat. 2; H. Stern, *op.cit.* (n.3) *passim*.

36. E.M. Wightman, *op.cit.* (n.14), 170 Pl. XIII.

37. J.M.C. Toynbee, *op.cit.* (n.24), 161, no. 89.

38. G. Henderson, *Early Medieval* (Harmondsworth, 1972).

39. J.M.C. Toynbee, *op.cit.* (n.24), 184 and Pls. CL, CLI No. 138 = RCHM Eburacum 141ff. Pl. LXVIII Nos. H. 2442, H. 2444.

40. G.M.A. Richter, *Engraved Gems of the Romans* (London, 1971), 120ff.

41. H. Webb, 'The Reign and Coinage of Carausius', *Num. Chron.* fourth series 7 (1907), 1-88, 291-338, 373-426 Pl. III, 3 and 7.

42. J.M.C. Toynbee, *op.cit.* (n. 24), 189ff, Pls. CLXXVIII ff. No. 155

43. R.B. Bandinelli, *op.cit.* (n.3), passim.

44. M.I. Rostovtzeff, F.E. Brown, C.B. Welles, *The Excavations at Dura-Europos. Preliminary Reports on the 7th and 8th seasons* (New Haven, 1939), 326-369.

45. e.g. Munich, Bayerische Staatsbibliothek, Clm, 10291 f. 177, illustrated by. R. Goodburn and P. Bartholomew *op. cit.* (n. 16), Pl. XXI.

46. *ibid.*, frontispiece, where three of the medieval MS versions of the insignia of the *consularis* of Campania are reproduced in colour.

47. *JRS* 53 (1963), Pls. XI ff.

48. K. Weitzmann, *op. cit.* (n. 12), 108ff.

49. J.M.C. Toynbee, *op. cit.* (n. 1), 328ff, Pl. LXXVI.

50. K.S. Painter, *The Water Newton Early Christian Silver* (London, 1978). See now J. M. C. Toynbee 'A Londinium votive leaf or feather and its fellows' in eds. J. Bird, H. Chapman, J. Clark, *Collectanea Londiniensia Studies presented to Ralph Merrifield* (London 1978).

51. *Id.* 'A Late Roman Silver Ingot from Kent', *Ant. J.* 52 (1972), 84-92.

52. J.M.C. Toynbee, *op. cit.* (n. 1), 311.

53. *Transactions of the Birmingham and Warwickshire Archaeological Society* 85 (1972), 157.

54. O.A.W. Dilke, *op. cit.* (n. 5), 159-177.

55. *Ant. J.* 54 (1974), 117ff. Also *cf.* K. Weitzmann, *Late Antique and Early Christian Book Illumination* (London, 1977), 22, and Pls. 11-14. He concludes that the style is probably Western and that the manuscript may have been produced in Gaul during the second half of the fifth century.

PLATE II. a. Shepherds. Vatican, Biblioteca Apostolica Vaticana, Vat. Lat. 3867, f. 11r

PLATE II. b. Turnus. Vatican, Biblioteca Apostolica Vaticana, Vat. Lat. 3867, f. 74v

PLATE III. a. and b. Scenes from the Aeneid, mosaic, Low Ham Somerset, with detail below. Photographs: Warburg Institute, London.

PLATE IV. Filocalus' Calendar, Dedication, Vatican, Biblioteca Vaticana, Barb. Lat. 2154, f. 1r

PLATE V. Filocalus' Calendar, February, Vatican, Biblioteca Vaticana, Barb. Lat. 2154, f. 17r

PLATE VI. Filocalus' Calendar, Alexandria, Vatican, Biblioteca Vaticana, Barb. Lat. 2154, f. $3^r$

PLATE VII. Africa, Oxford, Bodleian Library, MS. Canon. Misc. 378, f. 147

PLATE VIII. a. Head of Christ, mosaic, Hinton St. Mary, London, British Museum

PLATE VIII. b. The poet Corydon, Vatican, Biblioteca Vaticana, Vat. Lat. 3867, f. $3^v$

PLATE IX. Sculpture, Murrell Hill, Cumberland.
Photo: Warburg Institute, London

# HARVESTING MACHINES, PALLADIUS AND TECHNOLOGY IN THE LATER ROMAN EMPIRE

by K.D. White

The Anonymous author of the De Rebus Bellicis consistently claimed that his military inventions had the great advantage of being economical in manpower (Pref. 12, 13, 14; XIV 2; XVIII 4, 11; cf. the changes in the system of recruitment detailed in V 7 intended to fill the depleted ranks of the army). Indeed, it would not be too much to say that he saw the slavation of the empire largely in terms of technological innovation. In one area, and to one class of people, his message will have struck a familiar chord: the wealthy landowners of Northern Gaul had long used a labour-saving device, the vallus, in harvesting. This remarkable machine, or rather a heavier version derived from it, was cited by Thompson (p. 80) in his discussion of the level of technological competence of which the Romans of the Anonymus' day were capable. Since Thompson wrote, however, the archaeological evidence for the appearance of the vallus has grown (Pls. Xa, b, Fig. 1). The machine is now seen to bear an uncanny resemblance to one of the Anonymus' actual inventions, the Tichodifrus (Part 2, Pls. VII & VIII). For a number of reasons therefore, it will be useful to look again at the evidence for the use of reaping machines in the Late Empire and for this we must turn to Palladius, and his work, the De Le Rustica.

## 1. The Date and background of Palladius

The date of this work, the last of a remarkably rich and well-assorted corpus of writings on Roman agriculture, cannot be fixed by any specific internal evidence, which tells us only that he owned landed property in several different areas. If the author of the De Re Rustica is to be identified with the Gallic friend of the fifth century poet Rutilius Namatianus, the work must belong to the fifth century, and be dated not earlier than AD 450[1], that is to say somewhat later in date than the composition of the De Rebus Bellicis. There is no internal evidence to support the guess that Gaul was the author's homeland, but the Tables of hours of sunshine, which are appended to each of the monthly accounts of operations which comprise the text, are evidently based on sundial readings for lat. 45°, which is consistent with that conjecture[2].

## 2. Technology in Palladius

The author's recommendations on the various operations of the farm are, for the most part, condensed versions of earlier writings, notably those of Columella. But there are two passages containing material of much technical interest, which are derived from independent sources, probably original to the author, and reflecting his own personal observation. These comprise (1) a discussion of building materials and techniques which evidently reflect

PLATE X. _a_. and _b_. Front portion of a _vallus_, relief Buzenol (_a_)
Rear portion of _vallus_, relief Arlon (_b_)
Copyright A.C.L. Bruxelles

FIG. 1 Reconstruction of the *vallus*, based on the reliefs from Buzenol and Arlon (cf. Pls. X. *a*. and *b*.)

contemporary architectural practice, more particularly as concerns the construction of cisterns;[3] (2) a chapter on harvesting of wheat which stands at the head of the Calendar for the month of August, and which is entirely given over to a detailed account of an animal-operated machine for harvesting grain, the text of which runs as follows: 'On the more level plains of the Gallic provinces they employ the following short-cut or labour-saving device (compendium) for harvesting. With the aid of a single ox, the machine outstrips the efforts of labourers and cuts down the time of the entire harvesting operation. They build a cart, mounted on two small wheels; the square surface of the cart is made up of planks, which slope outwards from the bottom, and thus provide a larger space at the top. The height of the planks is lower at the front of this container (carpentum); at this point a large number of teeth with spaces between them are set up in line to match the height of the ears: the tips of these teeth are upturned. Attached to the rear of the vehicle are two very small yoke-beams, resembling the poles of a litter. At this point an ox is harnessed by means of a yoke and chains, with his head pointing towards the cart. He must be docile, so that he will not exceed the pace set by the driver. When the latter begins to drive the cart through the standing corn, all the ears are seized by the teeth, and piled up in the cart, leaving the straw cut off in the field, the varying height of the cut being regulated from time to time by the cowherd who walks behind. In this way, after a few journeys up and down the field, the entire harvesting operation is completed in the space of a few hours. This machine is effective on open plains or where the ground is level, and in areas where the straw has no economic value'. Palladius De Re Rust. 7.2.2.

## Discussion

The above account of the machine, which is very precise on matters of critical importance for its efficient operation (e.g. the recurved tips of the teeth), suggests very strongly either that the author had seen the machine in action, or that he he derived his account from someone who knew it at first hand. It is therefore surprising to find modern writers on ancient technology either playing it down[4] or denying that it was ever used.[5] Although the machine here described by Palladius has no precise parallel in our other sources of information, which are almost enitrely archaeological, we have no fewer than five identifiable representations of machines belonging to the same genus as Palladius' harvester, but to a different species, the vallus (Pl.X; Fig. 1); four of these are funerary reliefs, and the fifth is part of a sculptured representation of the Seasons.[6] Investigation of these monuments shows that all conform to a basic design, that of a simple wooden frame mounted on a single axle, the front portion being fitted with a row of teeth, exactly like those of Palladius' machine. These lighter specimens were similarly propelled from behind by an animal in shafts. But there the resemblances end. Whereas the tumbril-type body of Palladius' harvester is itself the receptacle for the heads of grain, the others have either an underslung container or a sort of grass-box behind the teeth; in either case the capacity is much more restricted, involving frequent stoppages to empty the container. The difference in size and weight is further emphasized by the animal: Palladius' heavier machine is pushed by an ox, the others by a mule or donkey. Since the earliest and latest references to these machines are separated by a time-span of approximately four centuries (from c. 50 AD to c 450, improvements in design would be likely,[7] and Palladius' reaper seems to indicate just such an improvement.

## Technical and developmental problems

While there is evidence of technical development in the substitution of mechanical for manual methods of harvesting grain, the literary references, coupled with the monumental evidence that has so far come to light, show that for some reason the machines did not spread beyond the area of their invention, namely the trunk road which linked the city of Reims (Durocortorum), the capital of the Remi in what is now Champagne, to Trier (Augusta Treverorum) the capital of the Treveri, and for some distance to the south of that road. The area comrpises undulating downland, very suitable for the large fields which would favour mechanical harvesting. That we find no evidence of their use in Italy is not surprising, since the combination of sown and planted crops which was standard in the Peninsula in Roman times from the second century B.C.[8] would preclude the use of machines requiring flat, unimpeded stretches of open ground. But the failure of the invention to spread elsewhere was probably due to a variety of factors, three of which are spelt out by Palladius in the concluding paragraph of his account: the machine, he says, is only suitable on ground that is open and level, and where the straw has no economic value. To take the first two points: open, level ground was not too plentiful in the north-western provinces: the Ardennes forest, which lies close to the area of the invention, is succeeded to the northward by the extensive lowlands of Flanders and Holland, which were subject to repeated inundation from the uncontrolled waterways that constituted the lower reaches of the Meuse and the Scheldt. To the westward, much of northern Gaul was heavily forested in antiquity, while to the east lay the vast forests of free Germany beyond the Rhine frontier. Most of the surrounding regions were thus unsuitable for the spread of the invention. Turning now to the major wheat-producing areas of the Empire, Egypt is obviously ruled out. Both Sicily and North Africa will have provided suitable terrain, but in these regions, as over most of the lands bordering the Mediterranean, the straw, which was destroyed by trampling, was of considerble economic importance, both as roofing material and as bedding for stock.

## Time-saving and manpower economies

The introduction of machines, as Palladius points out, meant savings of both time and manpower. In the north west, speed was—and is—of the essence of the harvesting operation. A whole crop can be ruined by storms or persistent rain. This factor was obviously much less important in the Mediterranean regions. Manpower questions, for lack of specific evidence, are largely matters of speculation, some of it wild and most of it unprofitable.[9] Seasonal shortages of farm labour, particularly at harvest and vintage time, are mentioned from time to time in the literary sources, but such shortages are common in all periods where manual methods predominate: in antiquity, as in later times, these were met by employing itinerant labour. There is, in fact, no evidence of any significant shortage of manpower right down to the later fourth century. But the steadily increasing prosperity of the regions adjacent to the Rhine frontier, for which there is good evidence in the neighbourhood of Trier, may have affected the local market for seasonal labour. Lack of incentive to technical improvement, however, is a factor to be considered. I have shown elsewhere[10] that the attitude of the landowning classes in Roman times was essentially that of the rentier; so long as they continued to obtain a steady return of around 5-6% on their investment, there was no motive for additional outlay designed to produce greater returns. This explains to some extent the contrast between technical progress in the military sphere and in certain industrial spheres such as mining, where the state provided the necessary funds, and the stagnation of the private sphere. From this point of view the restricted range of the Gallo-Roman harvesting machines stands in contrast with the development of _ballistae_ and other offensive weaponry.

In conclusion it can be said that under conditions of manpower shortage, the advantages of mechanisation _could_ be appreciated in Roman times, and the solutions of the Anonymus to this particular problem may have been along the right general lines, since the continued use of harvesting machines in Gaul down to the times of Palladius shows that the benefits of such labours saving devices must have been measured in hard economic terms. But to suppose that the Anonymus was inspired by the _vallus_ when he dreamt up his _tichodifrus_, would be to go beyond the evidence. This suggests that a heavier machine derived from the _vallus_, and not the _vallus_ itself, was employed at the time that the Anonymus was writing; but even if the light-weight _vallus_ continued in use, it is unlikely that the Anonymus could have seen it, for his own words (XVIII 5) imply that he came from a province where men thought of the Danube as _the_ river frontier, and the use of reaping machines never seems to have spread beyond the Rhineland and Northern Gaul.

Garden Flat,
6 Alexander Street,
LONDON W2

## NOTES AND REFERENCES

1. See J. Dalmasso in *Rev. Phil.* 43 (1915), 82ff.
2. See R. Harris in *AJP* 3 (1882), 411-421.
3. Discussed in detail by W. H. Plommer in his recent monograph, *Vitruvius and Later Roman Building Manuals* (Cambridge, 1973).
4. See e.g. M. I. Finley 'Technical innovation and economic progress in the ancient world', *Econ. Hist. Rev.* 2nd ser., 18 (1965), 29ff.
5. By C. Mosset, *The Ancient World at Work* (London, 1969) 35: 'It is doubtful whether Pliny's famous and controversial reaper was ever used'.
6. See K. D. White *Agricultural Implements of the Roman World* (Cambridge, 1967).
7. Denied, against the evidence, by Hodges, *Technology in the Ancient World* (Harmondsworth, 1971), 199.
8. See K. D. White *Roman Farming* (London, 1970), 47ff.
9. See the review of M. I. Finley of A. E. R. Boak's *Manpower Shortages and the Decline of the Roman Empire in the West* (Ann Arbor/London, 1955), in *JRS* 48 (1958), 156-164, showing that B. has antedated the shortage of manpower by two centuries.
10. K. D. White, *op. cit* (n. 8), 446ff.

# TAXATION AND SURVIVAL IN THE LATE FOURTH CENTURY: THE ANONYMUS' PROGRAMME OF ECONOMIC REFORMS

by Patsy Vanags

*Nobis perit deterior possessor effectus, nobis proficit non gravatus* (Val III Nov. 1,3,3.). 'A landholder rendered impoverished is lost to us; one who is not overburdened, profits us'. Thus wrote the emperor Valentinian III in his *novella* of 450 A.D., and the late Roman administration was more than aware of the truth of this simple precept. It is in this context, of the continuous fight against the adverse effects of taxation which the government was forced to pursue, that we can take another look at the Anonymus.

Essentially, the Anonymus argues that one of the major problems in the later Roman empire is the 'over-taxation' of the ordinary agricultural worker. It is an argument that has been restated by A. H. M. Jones.[1] For both, it is the *robor*, the well-being, of the farmer as tax-payer that is the key to the situation. At the time that the Anonymus was writing[2], tax-payers were sufficiently oppressed for him to emphasise that what was needed was to *restore* cultivators to their former strength–this was no warning of what *might* happen.

The Anonymus suggests two lines of approach to help the tax-payer. One (I.1) is to reduce the overall level of government expenditure, to ask for less in total: *Quae si providentia maiestatis imperatoriae reprimatur* (sc. the *profusa largitio* of the previous sentence)... *collatorum potius defecta subsidia recreantur*—'If the foresight of the imperial majesty restrain (this unbridled expenditure) ... the enfeebled resources of the tax-payer will be restored'. The other is to effectively increase the absolute number of tax-payers so that the burden of the taxes may be shared amongst more people and therefore reduce the load on each individual. This strategy is argued in chapter V, on 'Methods of economy in military expenditure', where he envisages (V.6) veteran soldiers retiring to cultivate the fields but paying taxes like everyone else: *Huiusmodi igitur provisionis utilitas in augmentum provincialium habet veteranos* ... *ad colendos agros praevalentes agricolas*, whereas in practice in the late fourth century veterans did not pay the poll-tax.

Both of these approaches would effectively reduce the tax-rate for all tax-payers. The late Roman tax system was based upon an assessment of the assets of an individual cultivator (amount and quality of land, amount of agricultural capital such as livestock, and labour inputs assessed sometimes on a separate schedule as poll-tax, allocating tax liability according to the resources of the individual.[3] The amount of tax demanded each year in principle altered with the needs of the government. The tax-rate levied on each standardised unit of account, or *iugum*, could vary according to whether the government n needed more revenue or less. In practice, the tax-rate escalated inexorably through the fourth century, as the government always decided that it was hard-pressed and needed more funds.

But a higher and higher tax-rate did not necessarily mean a bigger and bigger tax-yield. It could mean people refusing to pay at all—there was indeed a last straw which could be put on the camel's back. And here we can confront the Anonymus with the reality of the historical record. A dramatic reversal to the escalation of the tax-rate was engineered by the future emperor Julian. He brought in a lower tax-rate in Gaul in 365, reducing it from 25 solidi per caput (the Gallic version of the iugum) to 7 solidi, and the total revenue gained was more than had been brought in under the much higher rate.[4] That is, by asking for a feasible amount, there was a real chance of getting it. If too much was demanded, people would not pay up, and wauld fall into arrears, or join the brigands, or go over to the barbarians, according to their status and degree of desperation.[5]

We know that Valentinian and Valens continued the efforts to reduce the burden on tax-payers—Ammian praises both of them for this—the phrases he uses 'relieving the burden of the tribute' (of Valentinian), and 'relieving the weight of the tribute' (of Valens) stress the reduction of pressure on the tax-payer.[6] Themistius says in his panegyric to Valens that he reduced the overall amount of tax (i.e. the indiction) by a quarter.[7] In view of Julian's remarkable success in Gaul, it seems that more realistic management of taxation via a lower tax-rate, could result in better circumstances for tax-payers and better revenue prospects for the government. That is the irony of 'overtaxation'.

But why did the average cultivator suffer so much from the effects of taxation? Obviously, if the state were asking for too much, more than people could manage to pay, then life was hard for a farmer. Too much was also demanded of poorer people because the richer were able to avoid due payment in many cases. It may be indicative of the class interests of the Anonymus (for whatever he may have been he certainly wasn't a peasant) that he only tackles the general problem of overtaxation by the government and never mentions the evasion of taxes by the rich and moderately rich (which was the object of tough legislation by Valentinian and Valens). His graphic portrayal of the rich amassing great piles of gold and silver at the expense of the poor powerfully suggests economic inequality—but that is all.[8]

There is no hint either of the vicious connection between the rich staying rich and the whole system of administrative corruption that accompanied the assessing and raising of taxes. We will look at the Anonymus' treatment of the problems of administrative corruption in a moment. But if we have to choose, then the perks which might line the pockets of minor officials would be a good deal less costly in financial and social terms than the immunities purchased by the great landowners from officials higher up the administrative pyramid. It is the high level collusion between men of substance and officials which mutually benefits them—and does the greatest harm to tax-payers en masse because of the scale of evasion involved. We have only to recall that for the greater part of Valentinian and Valens' reigns the figure of Petronius Probus towers over the administration of the empire, Praetorian Prefect and head of the great house of the Anicii. Such a nexus of senatorial and political power must have thwarted many of the efforts to improve the fairness of the tax administration which were made at the time. To give the Anonymus his due, he may only have been able to cite 'the appalling greed of the provincial governors, which is ruinous to the taxpayers' interests' in such general terms (and even this is quite brave criticism).[9] To have said that the bureau of the praetorian prefecture was a nest of villains favouring friends and relatives was hardly possible when Petronius Probus' connections were so powerful.

Corruption amongst administrators is, of course, the evil against which every bureaucracy has to fight—the Anonymus' plea that the provinces should be governed by men who are 'honourable and lovers of honesty' casti et integritatis cupidi (IV,5.) is a pious if not naive hope. But his sharp words for governors who 'think they have been sent into the provinces as merchants' (velut mercatores) (IV,1.) are a much more pertinent comment which deserves consideration. Let us look at some of the circumstances of their 'mercantile activity' which the Anonymus considers to be so damaging to the public interest.

Professor A.H.M. Jones, in his commentary on this passage, suggested possibilities for administrators abusing such procedures as the purchasing of recruits, horses and grain, all of which the Anonymus cites as examples of corrupt practices.[10] What we should note here is the presence of the dilemma which existed for a provincial governor, corruption or no. In a sense, the whole complex of decisions which he was required to take as a local administrator committed him to behaving rather like a merchant—decisions on the price for recruits, the price for grain, whom to give building contracts to and at what rate. The fact of the matter is that provincial governors were deeply involved in local affairs and in the local economy: the extent to which the Roman imperial administration affected the economy as a whole was enormous. And most of the impact came through all the paraphernalia associated with the tax system.

The particular abuse which the Anonymus decries that highlights the whole complex of problems relating taxation to the economy, is coemptio, where a governor had to display the most mercantile of his talents.

Coemptio was the government purchase of grain, usually direct from producers but sometimes from merchants acting as intermediaries.[11] This procedure has always had a very bad rating by modern historians, not least because the chief description of how it might work is to be found in Procopius' Anecdota.[12] We would expect horrific stories of how the suffering cultivators were forced to provide grain at derisive prices in the Anecdota (which vilified Justinian and his administration), but the potential and actual evils of the procedure are also testified to by legislation regulating the abuse of coemptio.[13]

The real point is, that government purchase of grain could either harm or benefit the producer; it depended on the exact circumstances of its use. Clearly when coemptio was used judiciously, neither Procopius nor the legal documents would find it necessary to praise it.

The government purchase of grain has to be seen in the context of the late Roman tax system. Those producers growing significant amount of grain would have their potential for output assessed within the framework of the tax system. The levying of a specified amount of tax in grain in effect set a minimum target for production. When, because of the vagaries of the climate, output fell below the estimated level, tax payers would need a tax rebate, or fall into arrears of payment. Where coemptio comes in is when output was going to be better than the estimated average predicted by the tax inventory. The problem with setting a target for production was that overproduction could also be a real hazard for producers.

This mattered particularly when the government expected money payments of tax from grain producers. For in the event of a good harvest, the market for grain would become glutted, and prices would fall to levels where the net money income a farmer derived from a substantial crop would be less than he got when there was a shortage of grain, and prices were high. In this situation, where the government was responsible for committing producers to a certain level of output to try and cover themselves for a range of possibilities, it made sense for the government to step in and stabilize the grain market in the event of a glut. The buying-up of grain by the state (the biggest single source of demand, with vast purchasing power and with storage facilities) thus eased the ill-effects of a glutted market and helped farmers' incomes by preventing prices from falling to a low level.[14] The state of course served its own interests at the same time by creaming off a bigger surplus in good years, but coemptio also functioned as a stabilising mechanism in the local economy. That it could protect the robor of the taxpayers is clear from the fact that a government purchase of grain could be directly credited gainst the tax liability of an individual, allowing for the possibility of payment of arrears (or even future liabilities?) when an abundant harvest swelled output.[15]

Whether taxpayers benefitted or suffered from coemptio, then, depended entirely on the price the government was offering for grain relative to market prices. We should note that the government's intentions were, by the book, honourable; prices were meant to reflect the going market rate.[16] And we should also note that to assume, as many people have, that coemptio always implies compulsory purchase is a distortion of the meaning of both the Latin term and its Greek equivalent (συνωνή), which really only suggest 'buying up', a rather different matter.[17]

The best contemporary reflection of the many facets of government purchase of grain, and how it involved provincial officials as 'merchants' with very large possibilities for corrupt practices, comes from Africa. Ammian records that when Hymetius was governor in Carthage, collection point for the grain of Africa which fed Rome, he was reprimanded by Valentinian because he released stocks of government grain to the Carthaginians in time of famine; and then having made up the deficit when the harvest was good, did not remit sufficient funds to the treasury.[18] But exactly how Hymetius was at fault is not clear: did he keep some of the money himself, after paying either a reasonable price to farmers or a 'forced' low price to get a bigger cut for himself? Or alternatively, did he pay too high a price to producers for the grain, benefitting them (they may well have been large landowners to a significant extent) at the expense of the state treasury, which made less on the whole operation?[19] We have to leave the issue open—but one can see that an official even behaving 'with integrity' was treading a very thorny path indeed when the state actually required him to play the grain market like a merchant.

Let us now look at the Anonymus' proposals for reform in the conditions of military service and their economic implications. Firstly, the historical background.

In the mid-fourth century, the minimum length of military service that qualified a soldier for an honesta missio (honourable discharge) was twenty years, and to get full veteran privileges on retirement, he needed to stay another four years in the army. The retirement privileges given to soldiers by Constantine were: the grant of vacant lands, tax exemption in perpetuity; 25,000 folles in cash or a yoke of oxen and a hundred measures of seed corn. Under Valentinian and Valens, the grant of tax-free land was confirmed again-those with honourable or medical discharges were to get one pair of oxen and fifty measures of each kind of seed grain (wheat and barley); whilst ex-protectores (honorary or ex-members of the imperial bodyguard) got two yoke of oxen each and a hundred measures of each kind of grain.[20]

In chapter V of the treatise, the Anonymus puts forward his own suggestions for reforming the conditions of military service. His chief argument of economic significance is that the length of service should be shortened: this, he claims, will have two beneficial results—firstly, government expenditure on the army will be cut, because soldiers will be on the payroll for less time—secondly, since they will go into earlier retirement, they will still be physically viable as cultivators, and will be able to make fuller use of their retirement grants. Their role as active farmers should now be taken advantage of by the state, and they should be taxed, like any other provincials (V.6).

On the face of it, the Anonymus' suggestion that early-retired veterans should be given their capital grants to engage in agriculture and then pay taxes like everyone else, is an attractive one. Particularly since he is thinking of frontier areas, where land needed the hard labour only a younger man could give it, and where agricultural capital like oxen might be hard to come by, That it might be some years before such farming would pay off, he does not take into account.[21]

But the most important issue, which no doubt a late Roman administrator would have to take into account, is how far the giving out of retirement grants earlier, after a shorter period of military service, would be treated as a real improvement by prospective army recruits, if they then had to pay taxes like the rest of the provincials. Because at the time the Anonymus was writing, service in the army carried with it the privilege of tax exemptions for soldiers and their families, which were of real financial value, and must therefore have functioned as a genuine incentive to join the army in the first place.

The Anonymus' programme of reform would thus remove a powerful incentive for recruitment, and since he does not give actual figures for his proposed length of service or for the size of re rement grants, we cannot balance out the gains for a soldier of an earlier retirement against the losses of subsequently being taxed. What we can observe is that the actual priorities of the late Roman government were at variance with the Anonymus' idea—recruitment to the army was of paramount importance, and incentives were preferred to conscription. Valentinian and Valens stepped up the scale of tax allowances made to troops: e.g. under Constantine (325 A.D.), soldiers after 20 years service granted 2 capita remission then in 370 A.D. soldiers after 5 years service granted 2 capita remission while in 375 A.D. soldiers after 5 years service granted 4 capita remission (if comitatenses).[22]

The administration obviously felt that any loss of revenue through the giving out of such tax allowances was an acceptable cost to improve recruitment and to promote loyalty.[23]

There is a more complicated aspect of tax allowances to soldiers and their families, which concerns the economics of agriculture. We have seen that the Anonymus' scheme is to get soldiers back to the land as fast as possible—swords into ploughshares at top speed—to swell the labour force used in agriculture and hence to lighten the burden of taxation by spreading it over a larger number of cultivators.

Now what he assumes (in common with most historians who have studied this period) is that the more people there are in agriculture, and the fewer in 'non-productive' occupations like the army, the better. We cannot get very far with trying to weigh-up the value of agriculture against the value of the army, it is very difficult to assess the relative values of goods and services.[24] But we can explore some of the implications of the Anonymus' proposals, by seeing how they would affect the relative values of capital and labour in the rural economy.

The taking-away of recruits involves the removal of labour, which is replaced by capital, either in the form of new capital (the retirement grants, oxen and seed corn) or in the form of keeping existing capital that would usually have been taken away in tax (the tax exemptions). Was this a net gain or loss? To argue by analogy with most underdeveloped economies, in peasant families not suffering from extreme population pressure or the poorest of lands, not every pair of hands is fully utilised. The family unit uses as many people as happen to be in the family; the work-force is not planned to use labour intensively. When we talk of the recruits that Valentinian and Valens were inducing to join the army, we are talking of the young sons in peasant families. If there was a measure of underemployment in the work-force, the sending-off of younger sons was a loss that could be compensated for by the remaining workers. The steady recruitment of peasants from the Balkans, decade after decade in the late empire, without obvious adverse effects, bears out this hypothesis. While we cannot gauge results, it is worth remembering that the substitution of capital for labour in rural economies can even produce real economic growth.[25]

Thus the sending-off of an extra son may not have been a loss, in economic terms. The tax-exemptions which Valentinian and Valens actually gave out meant that the labour of the son was being exchanged for family income that would otherwise have gone as tax to the government. Of course, sustained benefit from such poll-tax exemptions would depend on the land-tax schedule of families not being increased to make up the difference given away in exemptions. This gets us deep into the complicated tax-game where what the government could lose of the roundabouts it might make up on the swings—revenue lost by remissions of poll-tax compensated for by increase in the land-tax.

Here the disappearance of the separate poll-tax assessment (perhaps even in the reign of Valentinian and Theodosius I) in the recruiting areas of the Balkans, and with it the whole incentive scheme of remissions, is highly interesting.[26] Why did the administration come round to a similar opinion as our anonymous writer, that soldiers, or their families, should be as liable for tax as other people?

I would suspect that one relevant consideration was the problem of administering poll-tax schedules efficiently. The detailed accounting that was needed to work out who was liable in each household, especially with exemptions being handed out, was work enough. But more important was the ease with which people could hide away to evade being listed. The early Byzantine administrators still wanted to measure labour inputs in their tax-schedules, but they counted each household for a hearth tax, the _kapnikon_, which replaced the poll-tax throughout the empire.[27] Farmhouses could not get up and hide away; it was better for a tax-assessor to cut his losses and do an estimate of the agricultural population that was not provoked into mobility at the whisper of a census.

So the less complications the better when it came to measuring people's liability for tax. But efficient measurement was only part of the problem in raising taxes. The Anonymus put his finger right on the central problem when he stated that it was the _robor_ of agricultural producers that was the key to survival. His scheme for making the capital grants of land and cash to soldiers on their retirement is similar to the grants given out under Valentinian and Valens: but since he suggests also a _shorter_ term for military service, his strong young veterans are setting up farms with very positive agricultural subsidies. We can see his point: arthritic old campaigners chasing a team of oxen round some semi-derelict field on the frontier is not a terribly good investment for other tax-payers' money. But while, traditionally at least, the Roman government was potentially ready to provide positive help to agriculture,[28] the policy which emerges with the late empire might be termed one of 'negative help'. Help was given—but against the ill-effects of the harm already caused by the government itself, largely through oppressive taxation.[29]

Taxation is always an unwelcome burden, but in the later Roman empire it hit the agricultural population particularly hard as more and more funds had to be found to counter the massive onslaught of the barbarians.

Taxation today drives some people to leave their native country, but probably drives few people to leave settled employment altogether. Neither does the promise to pay tax give us the right to work, or the right to property which will yield us an income. But in the late empire, people _were_ driven to defect to the barbarians, or to take up the life of brigands.[30] They were driven to contract away their freedom of movement and their bargaining rights against employers, in the form of serfdom known as the colonate;[31] they then accepted the chance to make a living by agreeing to pay tax in exchange for land. This was where the state stepped in to protect its weak but important members and its own revenue, by selling property-rights for the price of taxes.

Valentinian and Valens championed the cause of what were in effect state-sponsored squatters: if anyone settled on unoccupied land, farmed it and paid the taxes, they gained the property rights on that land. The previous owner was forbidden to arrive and behave as if he were landlord to the new occupants; specifically banned was a neatly-timed arrival for the harvest, to claim a share of the crop, the agraticum.[32] This granting of ownership of derelict land became available to anyone, though we can certainly imagine it applying to the veterans that the Anonymus mentions. Ironically their valiant conquests from the external enemy might then be threatened by the inroads of an internal enemy, who had owned the land before the barbarians came.

As with so many arrangements created in the late empire, it is easy to assume that none of these last-ditch efforts could have worked–which emperors could have legislated from a position of strength? But the laws of Valentinian and Valens were not mere mumblings of a worried administration. Their laws were used–ruthlessly and really effectively–against those, most notably the rich who were heavily behind with their tax-payments.[3] What is even more important, there were quite sound principles of public finance embodied in their laws, and these principles were at the root of the solid financial policies of the Byzantine empire. The economic arithemtic that lay behind the state taking the side of the squatter against a landlord-figure is simple: state and landlord were in competition for the peasant's surplus, the landlord would take it as rent, the state as tax. The public interest benefited more if it was tax. A peasant freeholder was of greater value to the state on the whole than a tenant-farmer.

So the ability to pay the tax on a piece of land continues to give access to the property rights over that same piece of land into the Byzantine empire, and remains as a cornerstone of fiscal policy. The individual squatter becomes powerful in the face of landlords if the weight of the state and the intricate machinery of a well-tried legal system are behind him. In the Byzantine East, serfdom does not come to dominate the economy, the peasant freeholder survives and prospers. Byzantine Egypt or Asia Minor never comes to resemble late Roman Gaul; the great landowners are only medium-sized compared with their Gallic counterparts and their power is checked by the survival of independent peasants.

When the right to land can be purchased through the payment of taxes, the exploited peasant has a chance of improving his lot so long as there is land for him to move to, and there was always plenty of land in an overall sense within the Roman and Byzantine territories. In a curious way, the later Roman empire thus evolved a mechanism for land resettlement which could achieve an ongoing process of land reform - a slow process of permanent agrarian revolution where the power of the great landlords was not broken at a blow by a Zapata, but intermittently as individual farmers came under pressure and moved off to start again with the backing of the state. This process effectively started under the emperors to whom the Anonymus was urging his programme of reform, it suggests a climate of optimism where bold solutions to problems were attempted.

Education Department,
The British Museum,
London. WC1B 3DG

## NOTES AND REFERENCES

1. A.H.M. Jones, 'Overtaxation and the decline of the Roman empire', *Antiquity*, 33 (1959), 39-43 and see arguments in the final chapter of *Later Roman Empire*.

2. As regards dating, I am assuming that the Anonymus is writing in the reign of Valentinian and Valens, in the light of Cameron's arguments, in 370 (or earlier). The earlier one chooses to place the Anonymus, the more it might appear that his advice in economic matters was taken!

3. The best short account of how the tax-rating was organised is in Jones *Later Roman Empire*, 61-8.

4. Amm. xvi.5.14.

5. One of the largest recent bibliographies on the late Roman tax system can be assembled from the paper by A. Bernardi, 'The Economic problems of the Roman empire at the time of its Decline', chapter I in C. Cipolla, (ed.) *The Economic Decline of Empires*. (London, 1970)

6. Amm. xxx.9.1; xxxi, 14.2.
   see also W. Goffart, *Caput and Colonate: Towards a history of late Roman taxation* (Toronto, 1974).

7. Them. *Or* viii.113a, τῷ τετάρτῳ δὲ τὴν ἴσην μοῖραν παρελύσατε τῆς βαρύτητος.
   I assume that this means that taxation had been reduced by a quarter, as does J. Karayannopoulos, *Das Finanzwesen des frühbyzantinischen Staates* (Munich 1958), 194. cf. Jones, Later Roman Empire, p. 147 who takes it to be a reduction of a half.

8. II.4. *Ex hac auri copia privatae potentium repletae domus, in perniciem pauperum clariores effectae, tenuioribus videlicet violentia oppressis.* The *tenuiores* are the ordinary poeple, 'la majeure partie de la population de l'empire' (J. Gage, *Les classes sociales dans l'empire Romain* (Paris, 1964), 283)

9. IV, 1. ....*iudicum execranda cupiditas collatorum utilitatibus inimica*.

10. Jones *op.cit*. (n. 1) 39-40.

11. On *coemptio*, see E. Stein, *Histoire du Bas empire* (Bruges, Paris, 1959, 1949) vol. 2, 199-203, Jones *Later Roman Empire* 460-1, 839-41, 844, 864; S. Mazzarino *Aspetti sociali del quarto secolo*, (Rome 1951) 55ff, particularly for its relationship to the commutation of 'in kind' levies to cash, *adaeratio*.

12. Proc. *HA*, xxiii, 9-14.

13. See n. 11 *supra*.

14. More on the economics of government grain purchases in Vanags, *Econ. Hist. Rev.* forthcoming.

15. *Coemptio* credited against producers tax liability, *CJ* X, 27, 2. Ὑπολογυζέσθ δὲ τοὺς πιπράσκουσι τὰ τυμήματα τῶν εἰδῶν εἰς τὰ συντελούμε παρ' αὐτῶν ἐν χρυσῷ δημόσια.

16. *Corpus Iuris Civilis*, ed. Krueger *et al.* (Berlin 1928-9) 3, App. 7. *Constitutio Pragmatica*, 18: ... *pretiis videlicet pro specierem venalitate, quae tunc temporis in foro rerum venalium obtinere noscuntur. Statuendis, ipsis tamen pretiis specierem unicuique collatorum in tributorum imputandis exactionem*. There was no way, of course, in which legislation could cover all the eventualities of changes in the price level of the open market; clearly, the price of grain could go up, however glutted the market, once people were confident that the government was going to step in and offset excess supply. The presence of οἱ πατέρες τῶν πόλεων was recommended during the negotiations for grain purchase, to protect local interests (*CJ*, X, 27, 12).

17. Cf. entries in Lewis and Short under *coemptio* and Liddell-Scott-Jones under συνωνή .

18. Amm. xxviii.1.7. *Cum [sc Hymetius] pro consule regeret, Carthaginiensibus victus inopia iam lassatis, ex horreis Romano populo destinatis, frument um venumdedit, pauloque postea, cum provenisset segetum copia, integre sine ulla restituit mora. Verum quoniam denis modiis singulis solidis indigentibus venundatis, emerat ipse tricenos, interpretii compendium ad principis aerarium misit. Ideoque Valentinianus, per nundinationem suspicatus parum quam oportuerat missum, eum bonorum parte multavit.*

19. Cf. F. Dölger in *Byz. Zeitschr.* 34 (1934), 71, 'Es ist mir wahrscheinlich, dass die συνωνή von Anfang an nur von den Grossbesitzern erhoben wurde.'

20. Retirement grants to veterans: *CTh.* VII.20.3 (320, 326): *CTH* VII.20.8. (364).

21. The late Roman government did take into account the time-lag operative between investing capital and effort in land and getting substantial results. Lands that required many years of effort before a decent level of production could be achieved were rated sympathetically in terms of the rent and tax demanded, usually on a share-cropping or tithe basis that would adjust the amount taken automatically as production went up. The giving out of tax-exempt lands to veterans might not be such a big concession after all, if those lands would not have a substantial yield for sometime.

22. *CTh* VII.20.3 (325): *CTh* VII.13.6.(370); *CTh* VII.13.7. (375).

23. *CTh* VII.13, 7(3), *Ipsorum etiam, qui militaturi sunt, privilegiis accedentibus facilius devotio provocatur,...* In the same law the administration demonstrates its concern for the loss of revenue such tax-exemptions might involve–*Et quia publica utilitas quoque cogitanda est, ne sub hac indulgentia insertae capitationis numerus minnatur, ex*

incensitis adque adcrescentibus in eorum locum, qui defensi militia fuerint, alios praecipimus subrogari. The devotio which hopefully resulted from such privileges might be construed as being personal loyalty to the emperors, bought in an analogous way to the purchase of loyalty with donatives. But since the veterans in the Balkans had been sufficiently disloyal as to join the rebellion of Procopius in the reign of Valens. (Amm. xxvi, 7, 1), along with currently-serving soldiers, longer-term privileges, of continuing economic value, might buy more devotion.

24. E.g. Jones, who is not alone in his notion of the large army and bureaucracy being parasitical on the economy. Cf. Later Roman Empire, 1045, 'the basic economic weakness of the empire was that too few producers supported too many idle mouths'. While the Anonymus does not complain about there being too many civil servants, any distinction between civil and military officials is blurred because soldiers assisted in administrative matters.

25. A major issue of modern development economics. For a recent example of the beneficial effect on peasant incomes of capital/labour substitution as a result of the recruitment of peasants into the army, see the analysis of the impact of the Vietnam war on the local economy, in Sansom, R. L., The Economics of Insurgency in the Mekong Delta of Vietnam, (M.I.T. Press, 1970).

26. Poll tax abolished in Illyricum, CJ XI. 53.1,(371): in Thrace, CJ. XI. 52.1. In Bithynia and Pontica it was reduced by a quarter by Anastasius, Joh. Lydus Mag. iii.47.

27. On the kapnikon, and on Byzantine taxation generally, see the chapter by Ostrogorsky, 'Agrarian conditions in the Byzantine Empire in the Middle Ages' in CEHE, vol. 1.

28. For example, the project to extend the cultivation of the vine attributed to the emperor Probus, in SHA, Vit. Prob. 18.

29. Poor taxpayers were treated more leniently with outstanding tax-debts than the rich, for example. CTh, XI. 1.27 (400/5).

30. Brigandage–among veterans themselves–CTh VII. 20.7. (353): defection to the enemy, Them. Or viii. 115a.

31. The institution of the colonate involved limitations on the mobility of the peasants and curtailment of their liberty: see Jones, Later Roman Empire.

32. CTh vii. 20.11 (368), a general provision for veterans to cultivate derelict lands, and the forbidding of former owners from claiming the agraticum. CJ XI. 59.8 (386), remission of arrears offered to owners of deserted land, or, if they failed to return, the offering of the land to anyone who was prepared to pay the taxes on it.

33. Valens' father-in-law, Probus, tracked down long-overdue tax-debts in the Balkan provinces, subjecting even the rich and powerful to ruthless investigation and torture, Amm. Marc. xxvi.6.7, 365 AD.

## THE ANONYMUS : A NUMISMATIC COMMENTARY

by Richard Reece

The references to coins and monetary matters in the Anonymus are so much a part of the general argument of the tract that they need close scrutiny and detailed exegesis. At present I can do no more than suggest some ideas rather as an interim note than a final report.

First and foremost, the Anonymus stands head and shoulders above other writers of the Roman Empire because he writes not only about denominations and designs of coins, but also about the day to day theory of coinage. Livy, Pliny, Tacitus, Dio and other historians mention odd facts about coins but they never take up the subject of coinage as such. They seldom give any reasoning behind any changes and avoid comment from their own standpoint on the facts that they give us. The Anonymus wades into economic debate with the express purpose of detailing past events in coinage to prove his point. He is not quite alone in this, but his companion is perhaps even more difficult to identify, date and evaluate. It is inevitable that a Symposium on a late fourth century topic should involve the writer of the *Historia Augusta*; I make no apologies for introducing him in the singular for the coin references in all the *Vitae* of the *Historia Augusta* are so eccentrically uniform that I can only believe in one writer.[1]

Like many writers before him, but in a more detailed fashion, the Anonymus is worried (I.1ff.) about the squandering of gold and silver by *largitio immoderata*. He exhorts the emperor (II.7), *repressa largitate*, and all will be well. The Scriptor has a similar matter on his mind. He places it in the reign of Severus Alexander[2] who, we are told, called in the gold coins of Elagabalus which had been two, three, four or ten times the weight of an aureus, or even up to a libra.[3] This comment serves to put the Scriptor on a level of coin reporting equal with the Anonymus but it is the reason for this change that is of extra interest; "Alexander" said that if you have coins which are ten times a gold piece or more, and you hand out a few gold coins, you give away ten times as much as you ought.[4] The remedy was to strike sub-divisions instead of multiples, first halves, then thirds and finally quarters, because he could not produce smaller coins, but as taxes did not come down as far as expected he had the (halves? and) quarters melted down and just struck *tremisses* and *solidi*.[5] *Tremisses* and *solidi* are flagrant anachronisms for someone writing about Severus Alexander and are difficult for someone writing under Constantine; they belong far better in the world of the Anonymus. At a stringent minimum we can say in safety that two writers in the age of the *solidus* were concerned with unnecessarily generous largesse, the forms and denominations of the coins, and the depletion of stocks of precious metals. Unfortunately this parallel may demote the Anonymus in my estimation for while the *Historia Augusta* loses all point if it was not intended for publication and must, therefore, be regarded as conforming to some literary conventions, our author is writing only a pamphlet which was probably never meant for public consumption. Mentions of coins may well have been made in the thousands of imperial memoranda which have perished, so the Anonymus' main virtue may be no more than survival. The historical flashback which the Anonymus gives us is a subject for antiquarian rather than numismatic enquiry because the author must be relying on textual rather than material evidence. One illustration of the traditions in general circulation, amongst the many which could be picked up, comes from another fund of monetary information, the *Satyricon* of Petronius, where in section 88 we are told that even the Senate often promise 1000 librae of gold to the Capitoline temple; these gifts to Jupiter mean that no one need be ashamed of offerings of money. This is the Anonymus' second phase, when bronze was abundant, rejected for ornamenting public buildings, and coined in such a way that it was worth its own weight (*aes validum ipso pondere* I. 8). The coinage had previously been instituted for two reasons - as a means of regal grants, which may be a tactful way of saying "paying the royal bills", and for everyday commerce (*emendi et vendendi utilitas* I.6). Michael Crawford[6] has questioned this second use of coinage and has documented his criticisms for the late Republic and early Empire, but neither his criticisms nor his evidence extend much beyond the early principate; economy and the use of coins after A.D. 200 may well not obey the more classical rules he suggests.[7]

Bronze had become more common and was minted as gifts for the troops and for private transactions (I. 9). This, if his history is indeed as much in chronological order as it seems to be, is the first century A.D., when the *sestertii* of Gaius and Nero are superb artistic productions, sometimes with purely military types such as the *adlocutio*, which may have been meant for official donatives rather than everyday commerce. For buying and selling it seems very likely that *assess* and *dupondii* were supplied, certainly to the troops, and probably to commerce in general.

The illustrations at the end of Chapter I (Part 2, Pls. I and II), do not, in their present form, fulfill the author's intention - to illustrate his history so that it might be more easily followed. The discs "*de corio*" are presumably of leather with a square of gold leaf at the top; they correspond to the description but give us no further help. Gold and clay coins must remain a mystery until the *Codex Spirensis* reappears, or the Carolingian copy comes to light. They present either a muddle, or a purely Renaissance style which defies analysis. It could be argued that the bare head and hairstyle of wavy locks in the Oxford *MS.* resemble those of Augustus;[8] that the laurel wreath and upward straining portrait owe something to Constantine[9] but I put these parallels forward without conviction.

Chapter II poses severe problems. The first sentence tells us that under Constantine generous largesses made gold, formerly very valuable, take the place of bronze in everyday commerce. Silver, quite rightly, for "*Constantini temporibus*" (or is it an accident stemming from the date of our author?), is not mentioned. This is archaeologically unacceptable. The coins commonly lost on an ancient site must give some clue to the coins used in "*vilibus commerciis*". If, under Constantine, gold lost its value and was used as bronze had been used, we should surely find gold coins in archaeological contexts.

In dealing with hundreds of thousands of coins found on sites in Italy, France and Britain I can never recollect seeing a gold coin of Constantine I. Either the Anonymus is using hyperbole here, or is simply wrong.

His explanation of the devaluation of gold is explicit and, as far as I can see, novel:- that which had been stored away in temples, gold, silver and lots of precious stones, was loosed, presumably via the mint, on the public. Since he has just spoken of the time of Constantine, this rush of precious metal is dated to his reign. Here, Mr. Stevens has given to us with one hand, but taken away with the other. He has provided Professor Thompson with a footnote (p. 32 n.2) to the effect that the reign of Constantine was characterised by an increased emission of gold coinage - and has not substantiated the claim; this is the debit: but he has given us the passage from the *de Errore Profanarum Religionum* of Firmicus Maternus[10] in which Constantius II and Constans are implored, or instructed, to cook the gods; for this his balance moves much in credit. This apparently Christian advice seems less bigoted when the mention of exactly the same course of action recorded by Valerius Maximus for the crisis of 82 B.C.[11] is remembered. Statues, even of gods, could often be reckoned as disposable assets. However, Mr. Stevens has moved us on to the joint reigns of the sons of Constantine which ended when Constantius II ruled alone from 350. The reason that I am so anxious to move later in the fourth century is purely archaeological and needs brief digression.

There are at least two ways in which an issue of coins may be judged large or small. The first method concerns survival. If many coins of an issue survive on sites and in collections, the issue may be called large, a quick way of checking is by counting the numbers in rich collections such as that of the British Museum. If few coins of an issue survive that issue may not be called small because the coins may have been hoarded, melted down, or recalled after issue. It is reasonable to say that the issue might have been small. Secondly, some attempt at objective estimates may be made by examining all surviving coins in considerable detail distinguishing which came from different dies, and coining up the number of dies known. If issue A has left twice as many dies per year as issue B it is a fair, but not foolproof, suggestion that issue A was twice as intensive as issue B. Few die studies have been done on late Roman coinage so I rely on rates of survival.

The great surge in coinage of precious metals seems to come soon after the *Tricennalia* of Constantius II in 353. If a museum or site is rich enough to have a few gold coins, it will typically have coins of Nero, Vespasian, perhaps the Antonines, and then a gap which is invariably closed by an issue of Constantius II recording *VOT/XXX/MVLT/XXXX*;[12] after this several gold coins may follow. The same is true in silver. I gather from Dr. Kent that the forthcoming Vol. VIII of *Roman Imperial Coinage* may tie these large issues to Constantius' visit to Rome with a date of *c.* 357-61. Since we have this sudden emission of precious coinage, I prefer to bring the account of the Anonymus into accord with archaeological fact rather than to leave it in textual isolation. This would then mean that we can accept the Anonymus' account by separating the confiscation of gold and the issue of it as coinage, attributing the former to Constantine and the latter to his son. The temptation to make a tiny emendation so that section II.1 begins "*Constantii temporibus...*" should almost certainly be resisted because our author is talking about the time of confiscation of precious metal as well as its issue, and for this "Constantini temporibus..." seems to be correct.

Chapter II posed intriguing problems which could be resolved by simple rearrangement and the recognition of hyperbole. Chapter III is of much sterner stuff. The basic problem concerns the fashioning (figura) of solidi which is depravata (III.1). We know exactly why this is a sad state of affairs - the image of royal majesty is diminished because the coins bearing it are refused, and this is the fault of the mint. Solidi are so uncertain that downright cunning has entered into any commerce where they change hands and no straightforward simple dealing is possible. We know the cure - put all the mint workers on an island, out of contact with the public, out of trade, they will then have no motive for fraud, and all will be well (III. 2, 3). If we know the effect and the remedy it may seem pedantic to worry about the precise nature of the fraud.

Professor Thompson settles on a perfectly good meaning of depravata, "debased". A literal rendering of "depraved" conjures up magnificent visions of the Fall of the Roman Empire, but does not help, and "malformed", which is near the original meaning of physical abnormality, begs the question of whether it is shape or substance which is below standard. I prefer "distorted" leading to a translation of "irregular". While Thompson has marshalled references to debased solidi[13] it is very hard to find examples of such coins, though plated gold and silver coins of the late fourth century are fairly well known, perhaps especially from Britain. To give specific examples there are among about 6,000 coins in the Corinium Museum, Cirencester, plated silver coins of Julian, and a solidus of Valentinian I with gold plate on a copper core. Erratic coins with reversed letters or odd style do occur (Bourges Museum, France) and it may be these which the workers in the mint produced unofficially. Most of the references to solidi in the Theodosian Code concern haggling over weight so that I feel that the issue of debasement should be left open. Thompson has noted the correspondence of language betwen the coin details of the Anonymus and the Scriptor, and this adds confirmation to my earlier suggestions.[14]

Finally, we come to the illustration at the end of Chapter III (Part 2, Pls. III, IV), which the author tells us is to make clear the nature of future, improved, issues. We are given only pictures of gold and bronze. Earlier (III.2) the reform of the mint is said to concern solidi and nummarii. The Theodosian Code uses nummus as a specific coin or unit of reckoning - perhaps the ubiquitous small copper coin of the late fourth and fifth centuries was a five-nummus piece, and this might encourage us to take nummus and nummarius as copper coins. The Historia Augusta, however, scatters nummi of gold, silver and bronze through its pages[15] so we ought perhaps not to take the Anonymus' nummarii automatically as specifically bronze. However, the picture with the attendant phrase, Tam aereae quam aureae, (III.4), makes it quite clear that silver is missed out. This is true of the whole tract - silver is a precious metal like gold but it never marches by itself and never explicitly produces coins. The Anonymus might almost have been living in a world devoid of silver coinage.

In the fourth century there was always, in theory, a silver coinage, but silver of high purity in good supply is confined to the years 357 to c.450 or even later in the East. I have already argued that the flood of precious coinage

FIG. 2 (above) Line drawing of the illustration in 'C' of proposed coin types (cf. Part 2, Plate III. a.).
Drawing by Simon James.

FIG. 3 (below) Models used by the illustrator of 'C'.
RIC Aug. 370, Dom. 270, Nero 328
Drawing by Simon James

starting in 357 is known to the Anonymus, this means that if his world really is without a good supply of silver, it must be after c. 420 in the West or very late indeed in the East. This is at present a problem which may be due to nothing more than chance selection of coins for illustration.

The actual illustrations of three gold and three bronze coins bring me to the point where Mark Hassall first introduced me to this tract and the Oxford manuscript. It was his suspicion that he recognized the Altar to Rome and Augustus at Lyon on the left-hand bronze coin that gave us faith in the illustrations and made me identify the other types (Figs. 2 and 3). The left-hand coin is indeed that altar with only the legend ROM ET AVG omitted; as the other types belong to first century *asses*, it probably derives from an *as* of Tiberius[16] The centre bronze coin is an only slightly misunderstood type of Moneta such as the As of Domitian[17] Moneta has her scales as expected but the S of S C has been muddled in the scales; what should be a cornucopiae cradled in the left arm has been turned into an eagle held on the left hand. The legend has again been omitted. The right-hand bronze coin is copied from an *as* of Nero[18] showing a winged Victory flying left carrying a buckler inscribed SPQR. In this case, although the expected S C is very faint or even omitted, there was no legend on the original. The gold coins suggest the issues of Constantine with upward straining faces.[19] The centre coin shows a diademed head, the right-hand one a laureate head. This resumblance, however, cannot be taken to be as firm as the identifications of the bronze coins. It is virtually impossible that the gold coins should be the same size as Asses of the first century (c. 26 mm diameter); on the other hand, it is interesting that *solidi* and common bronze coins of the period 366-375 do have roughly the same diameter (AV c. 20 mm diameter, AE3 c. 19 mm diameter).

I would prefer not to attempt a comparison of the illustrations of the coins in different manuscripts until I can see the detail of each drawing in the original MSS. and in colour. All that can be said at the moment is that one illustrator copied six Roman coins. Either the illustrator of the Oxford MS ('C') faithfully copied a remarkably preserved copy which exactly transmitted the fourth century original, which means that the Anonymus was able to find, and chose to copy, three well preserved *asses* three hundred years old. Or, and I am afraid this is more likely, the Renaissance illustrator, seeing that he had to draw three coins, went to the patron's coin cabinet, found a portrait series of the Twelve Caesars, chose Tiberius, Nero and Domitian (room for speculation here!) turned them over and drew them.

Although the main ideas presented here are the same as those presented at the Symposium, they have been most usefully refined, amended and amplified in discussion with Andrew Burnett, the late C. E. Stevens, Dr. John Kent and Prof. Alan Cameron. Without Mark Hassall's interest in the tract and his great enthusiasm I should never have taken my present interest in the subject, and the Symposium would never have taken place. Together with all the other contributors I owe him a debt of gratitude for a welcome chance to air my ideas in a very helpfully knowledgeable company and convivial surroundings.

Institute of Archaeology,
31-34 Gordon Square,
LONDON WC1H OPY.

## NOTES AND REFERENCES

1. Robert Ireland and Ian Marriott came to the same conclusion from their linguistic study of the SHA, JRS forthcoming.
2. SHA Alex. Sev. 29.
3. ibid. 39.9.
4. ibid. 39.10.
5. ibid. 39.7.
6. M. Crawford 'Money and Exchange in the Roman World' JRS 60 (1970), 40-48. see p.40.
7. R. Reece 'Roman Coinage in the Western Empire' Britannia 4 (1973), 227-251.
8. RIC I Pl. II, No. 33.
9. RIC VII Pl. XIX, No. 105.
10. de Errore Profanarum Religionum quoted by C. E. Stevens in 'A Roman Author in North-West Britain' in $CW^2$ 50 (1951), 70-79, see p.75, n.26.
11. Valerius Maximus, Factorum ac Dictorum Memorablium Libri ix, vii. 6.4.
12. RIC forthcoming.
13. Thompson, 36 (n.2), citing C Th IX.21.2 and Lib. Or. XVIII.138.
14. Thompson, 36-7 (n.6): forma (III.4) Cf. SHA Tyr. Trig. 31.3: figuratio (III.4) Cf. figurare. SHA Alex. Sev. 25.9.
15. e.g. SHA Elag. 8.3; Tyr. Trig. 31.3.
16. RIC (Aug.), 370.
17. RIC (Dom.), 270.
18. RIC (Nero), 328.
19. RIC VII (Con.), 104-5.

# FRONTIER POLICY IN THE ANONYMUS

by Stephen Johnson

Two passages in the *de Rebus Bellicis* seem to refer specifically to frontier policy. In chapter V, the author is arguing for a cutback in military spending, and then, more determinedly, but in a related vein, in chapter XX he suggests a method of policing the frontiers with small *castella* placed at regular intervals of one mile, and paid for by the local land owners. By taking these passages in turn, it will be possible to consider what relevance the recommendations which the Anonymus makes have to standard Roman practice in the middle and later phases of the Imperial period. Most of the paper will deal with written sources, either from late Roman authors, or from inscriptions.

In the first passage, the Anonymus counsels early retirement for soliders, so that they do not become a burden to the taxpayer. This, coupled with rapid promotion will see a more rapid turnover of troops, which will ensure that there are always large numbers of experienced veterans still presumably in the prime of life. 'A provision of this sort will increase the population of the provinces by supplying veterans enriched with Imperial gifts who will be strong enough to cultivate the land. They will live upon the frontiers, they will plough the areas which they recently defended, and having won what they longed to obtain from their toil, they will become taxpayers instead of soldiers' (V.6. Thompson's translation).[1]

First, what is there about this scheme for settling retired soldiers on the frontiers which is novel? It had always been imperial policy to grant land to soldiers at their discharge.[2] On the face of it, this idea for settling veterans in frontier areas is a good one; but the indications from late antique sources are that it was already common practice by the fourth century. On discharge, the veteran would receive a substantial sum in cash, or an equivalent amount of land, with animals and slaves to work it. In Augustan times, land was plentiful, and thus came about the formation of many new veteran colonies throughout the empire, but later it became increasingly difficult to find enough land to give to everyone, except at the very fringes of the empire. Possibly, already in the second century, there was some system of veterans settled near forts, particularly in the Rhineland, where, in the legionary territory of Neuss there are a series of small *villae rusticae* which may have been connected with the farming of the *territorium legionis*: an ideal situation for the retired legionary.[3] He could till his fields, and yet not be too divorced from the military life to which he was accustomed. There is epigraphic evidence from Aquincum[4] and and Troesmis[5] recording *veterani et cives Romani consistentes ad canabas legionis...* showing that veterans might well stay on in the *canabae* of legionary or perhaps even anxiliary forts.

From the Historia Augusta, we can trace the deliberate policy of granting frontier land to soldiers and veterans at least as far back as Severus Alexander, who 'gave what land he captured from the barbarians to the commanders and soldiers of the limitanei with the stipulation that they should never belong to civilians. He added to this grant draught animals and slaves, so that the land bordering on barbarian territory should never be left vacant because of sparseness of population or the old age of the owners'.[6] Limitanei occur frequently in the sources; in a rescript in the Codex Justinianus dated to the early sixth century, Belisarius refers to land to be given to limitanei so that they can defend the forts and settlements of the frontier, and till the land, too.[7] The inalienability of such land given to these frontier guards is a well established fact, stretching back at least as far as the beginning of the third century.[8]

Clues as to the sort of people who were preferred as limitanei are given by a rescript of Honorius and Theodosius preserved in the Codex Theodosianus. 'Areas which have been granted to the gentiles by the benevolence of our predecessors for the guarding of the limes must be reserved only for men like those who were assigned the task previously. Otherwise, these areas are to be given either to gentiles, if they can be found, or at least, which is right and proper, to veterans.'[9] The policy of giving land near the frontier to barbarians themselves is also well established by the end of the third century Passages from the Panegyrici talk of Maximian and Constantius' grants of land to Franks, who had thereby a dual role of tilling the soil and bearing arms for Rome.[10] It is even possible, in parts of Gallia and Germania to distinguish, by a study of the types of hill-top fortifications, where some of these tribes were settled. Not all so called Fluchtburgen were used solely for that purpose, and some are remarkably similar to the kind of fortifications inside which tribes in free Germany were living at the same period.[11] The policy of deductio was seen as the developing answer to the problem of providing a jealous frontier defence and a buffer state between the unromanised barbari and those inside the empire.

Deliberate settlement of veterani on the frontiers would be considered at least as effective as that of barbari, for even though their term of service had ended, they were still in several recorded instances in the later period of Roman history a force to be reckoned with. In Ammianus alone, veterans releived the siege of Autun in 356 when regular troops proved useless;[12] they played an important part in the revolt of Procopius against Valens in 365;[13] and in 378, veterans formed a portion of Valens army against the Goths.[14] That such veteran settlements were set up in frontier areas is shown by the survival of inscriptions and texts which document overseers of these communities.[15]

In beginning to consider what kind of buildings such frontier settlement may have used, the other text from the Anonymus becomes of relevance. After the large scale veteran settlement of the first century, there were no more towns created as colonies for them to inhabit, so the type of buildings which may have housed these soldier-farmers if probably on a smaller scale, and somewhat like the small buildings which the Anonymus suggests should be put up all round the fringes of the empire. 'It is a good idea and beneficial to the state to take effective control of the frontier works which surround all the borders of the empire. Their safety will be better provided for by a continuous line of forts constructed at intervals of one mile with firm walls and very powerful towers. These fortifications should be constructed without public expenditure on the individual responsibility of the local landowners, with watches and pickets kept in them so that the peaceful provinces may be surrounded by a belt of defences and so remain unimpaired and at peace.' (XX, adapted from Thompson's translation).

This passage, with its intriguing reference to 'milecastles' has often been seen as a reference to Hadrian's Wall, and has posed since Camden the intriguing question whether the Anonymus visited north Britain.[16] But there is no mention of a wall as a continuous barrier to link these milecastles, ; rather, the frontier is to be understood as something almost separate from this system of castella. There is far more evidence to link this passage with practice on other frontiers, and in particular the provision of towers known as 'burgi' and 'centenaria'.

The development of the frontier in North Africa was described by Baradez as the gradual process of irrigating a strip of land 60 kilometres wide carved out of the desert, and patrolled or controlled from small burgi - blockhouses with several rooms which served both as farmsteads and as forts.[17] Similar systems to the one studied in such detail by Baradez have been found in other parts of North Africa,[18] e.g. Tripolitania,[19] Mauretania Tingitana,[20] as well as in Syria[21] and Judaea.[22] Their clear connection with a system of farming and a policy of frontier defence links these small towers closely with the disposition of limitanei, whether gentiles or veterans. In areas which are now relative desert, the traces of this frontier are still remarkably clear: elsewhere, and particularly in the Roman frontiers in Europe, the evidence is more complex.

Europe was in the main fortunate that most of her frontiers ran along the river lines of Rhine and Danube:[23] this gave a solid demarcation between the Empire and barbaricum, but even so, from early days, systems of small burgi began to be placed at strategic points along this defensive line. A series of inscriptions from Thrace records the construction of several frontier works under Antoninus Pius in 152-5.[24] One, from the region of Serdica, records that 4 praesidia, 12 burgi and 104 phruri were built ob tutelam provinciae Thraciae.[25] Another, in Greek, records the construction of πύργοι in another area of Thrace[26] and two others in Latin recorded similar defensive works at a similar date in other parts of the province.[27]

Elsewhere, too, there were similar measures at a date not too far distant from this. Inscriptions from sites in Germania Superior,[28] Pannonia,[29] Numidia,[30] and Mauretania[31] mention the construction of burgi under Commodus, Severus Alexander and Caracalla. But it is possible that all these were official frontier works, and not paid for by the local landowners themselves, as the Anonymus suggests.

There is also some evidence of the manning system used to keep pickets in such frontier towers. A long Greek inscription, set up at Pizus in Thrace in 202 records the establishment of a new emporion created out of nine former native villages. Among the incentives offered to those who joined this new community of their own accord were several of particular interest:' to ensure the greater prosperity of these emporia. I persuaded men of good standing to move in from the surrounding villages, and to induce others to move into the settlements stipulated that those who were desirous of doing this of their own accord should have the following great concessions by favour of the Augusti, that is, freedom from the levy of wheat for the city of Rome, and exemption from the obligation to furnish burgarii and garrison troops and provision for the imperial postal service.'[32] The privilege of being excused 'burgarii' duty suggests that local communities were expected to provide men to garrison frontier posts. In addition, a further clause in the inscription shows that the upkeep of the buildings of the new emporion was to be borne by the local decuriones themselves - perhaps more than a hint of the same suggestion which the Anonymus himself made with regard to the furnishing of his frontier castella.[33]

As early as the mid-second century, therefore there was already a system of watch-towers along at least some of the frontiers of the empire, a system which was to be manned by some form of conscription of local people. That this continued later is shown by a passage from the Anonymus Byzantinus, who in a passage of his περὶ στρατηγηματῶν entitled περὶ φρουρίων speaks of the of men who were requied for such duty. They must be of good character so that they are not tempted to desert, and preferably they should leave a secure family behind them to keep them loyal. Duty in shifts is recommended, and if the phrourion is exceptionally safe, there is no reason why a man should not also take his family with him, particularly if the post can be made self-supporting by a little gardening on the side.[34] Here the idea that settlers should be encouraged to go and live on the limes is very close. Further quotations from the Codex Theodosianus[35] support this interpretation of the way in which the duty of burgarii was organised.

But burgi also had a peaceful purpose. An inscription of 188 records a burgus set up specifically to watch over a market at a crossroads, burgum Commodianum speculatorium inter duas vias ad salutem commeantium.[36] Another inscription from Pannonia, of Valentinianic date records the construction of a burgus, cui nomen Commercium, qua causa et factus est.[37] Other burgi are mentioned in Spain, so they were clearly not only confined to the frontier areas, and their connection with the imperial postal system is also well established.[38]

The majority of these sites, particularly those of the later Roman period are relatively well known and published. Systems of frontier towers are known in Raetia[39] Maxima Sequanorum,[40] and Pannonia[41] and give some idea of the comprehensiveness of late and, in some cases, earlier Roman cover of these areas. Sites here are usually simple towers, sometimes surrounded by a ditch. About six have left building inscriptions, all dating to the joint reign of Valens and Valentinian, that is, between 367 and 372.[42] This tallies well with various literary sources, notably Ammianus Marcellinus, who in more than one place mentions Valentinian's schemes for fortifying the whole of the Rhine and Danube banks with such towers.[43] Among other emperors, only Diocletian has a comparable reputation for building.[44]

There are sites, however, which tally more with the description of fort and farmhouse given by the Anonymus. In Germania Superior a series of towers with external walls which also have external towers have been found, for example those at Asperden,[45] Heerlen,[46] Heumensoord,[47] and Hulsberg.[48]

But other small scale forts are known from elsewhere on the European frontiers, for example at Eisenberg,[49] Strasbourg St. Nicolas,[50] and Skt. Margarethen in Burgenland.[51] These possibly find a common ancestor in the burgus of Harlach[52] and other slightly earlier small forts in Gallia Belgica at Senon and St. Laurent.[53] Near Vienna, too, archaeologists have found what may have been the empty tomb of St. Severin at Heiligenstadt, which was in a building of 'burgus' type, and recalls the mention in Eugippius' Life of the Saint that he lived in a habitaculum quod burgum appelatur ab accolis.[54] All these last buildings in particular owe something to the fortified centenaria of the African limites.[55]

Perhaps the most interesting idea of the Anonymus is that his scheme should be financed by the landowners themselves. This, too, was not original. Several of the African cenenaria carry inscriptions showing that they were built by the owners of the surrounding estates.[56] There is as yet little evidence from elsewhere of the same situation: richer landowners may have provided refuges into which the local populace could flee in times of danger[57] or even have built for themselves walled villas defended by several outpost burgi.[58] As for the burgi closer to the limes itself, there is no inscription to record who paid for the towers erected, though they were likely to have been built out of public funds, and there is unfortunately no evidence to shed light on who lived in such little forts.

What, then, is the Anonymus' frontier policy? Broadly, it fits easily into the mainstream of Roman practice without contributing anything startlingly novel. It would be surprising if a writer of late fourth century date had any devastatingly new ideas about frontier control after Rome had had more than three hundred years' cumulative experience in the field. The policy of settling veterans on the frontier was, if not commonly practised, at least not unknown, and the Anonymus' recommendations about frequent milecastles had been common practice on the frontiers since the mid second century. If it is surprising that the Anonymus should counsel exactly what the Emperor was engaged in doing, then the date of writing should hardly be after 367, after which time it appears that Valentinian was most active in improving the European frontiers.

Inspectorate of Ancient Monuments
Fortress House,
23 Savile Row,
LONDON W1X 2AA.

## NOTES AND REFERENCES

1. On this passage see Thompson, 41-4, and Patsy Vanags (p. 51ff. above).
2. G. Webster, The Imperial Roman Army (London, 1969), 41-3, 257-8.
3. H. von Petrikovits, Das römische Rheinland Cologne and Opladen, 1960), 63 f.
4. CIL III.3505 (=ILS 473).
5. CIL III.6166 (=ILS 2474).

6. *SHA*, Severus Alex. 58. 4-5. Similar passages, for example *SHA* Probus 16. 6, may be a little more inventive, since there is no archaeological evidence that Probus ever actually held any land beyond the Rhine to give to *limitanei*. But similar settlements occurred later, see e.g. *CJ* XI. 60. 3.

7. *CJ* I. 27.2: '*limitanei, qui possint et castra et civitates limitis defenders et terras colere...*'

8. R. Macmullen, *Soldier and Civilian in the Later Roman Empire* (Cambridge, Mass., 1963), 12-13. Constraints were also exercised on sons, particularly those of veterans, to follow the same profession as their fathers. See *C Th* VII. 22.1.

9. *CTh* VII. 15. 1: *Terrarum spatia, quae gentilibus propter curam munitionemque limitis atque fossati antiquorum humana fuerit provisione concessa.... circa curam fossati tuitionemque limitis studio vel labore noverint serviendum ut illi quos huic operi antiquitas deputarat. Alioquin sciant haec spatia vel ad gentiles si potuerint inveniri, vel certe ad veteranos esse non inmerito transferenda, ut hoc provisione servata fossati limitisque nulla in parte timoris esse possit suspicio.* Dated to 29.4.409.

10. *Pan.Lat.* vii (vi) 6. 2. Zos. i, 46. Amm. xvii. 13.3.

11. See, for example the Glauberg, *Germania* 18 (1934) 297f., and, inside the Empire, the Odilienberg, *Germania* 36 (1958) 113f.

12. Amm. xvi. 2.1.

13. *ibid.* xxvi. 7.1.

14. *ibid.* xxxi. 12.1.

15. Macmullen, *op. cit.* (n. 8), 64, n.41.

16. Thompson, 72-3, the references there cited and in the bibliography s.n. Cadwallader Bates, Camden Stevens.

17. J. Baradez, *Fossatum Africae* (Paris, 1949), and 'Organisation militaire Romaine de l'Algerie antique et évolution du concept défensif de ses frontières', *Revue internationale d'histoire militaire*, 4 (1935) 25ff.

18. See B. H. Warmington *The North African Provinces from Diocletian to the Vandal Conquest* (Cambridge, 1954) 23-4.

19. R. Goodchild and J. Ward-Perkins, *JRS* 39 (1949) 81ff., 40 (1950) 30ff., R. Goodchild, *JRS* 43 (1953) 65ff.

20. M. Ponsich, *Recherches archéologiques a Tanger et dans sa region* (Paris, 1970) 343f.

21. A. Poidebard, *La trace de Rome dans le desert de Syrie* (Paris, 1934).

22. M. Gichon, *Studien zur Militärgrenzen Roms*, (Köln Graz, 1967) 175ff., and *Roman Frontier Studies, 1967* (Tel Aviv, 1971) 191ff.

23. Apart from periods in the first to third centuries when the area between Rhine Danube was protected by the continuous line of palisade and all its intermediate stages of development.

24. G. Mihailov, 'Le fortification de Thrace par Antonin le Pieux et Marc Aurèle', Studi Urbinati 35 (1961), 42ff.

25. The full inscription runs: IMP CAES T AEL HADRI(A)N ANTONINUS AUG PIUS PP TRIB/POT XV COS III PRAESIDIA/ET BURGOS OB TUTELAM PROVINC/THRACIAE FECIT CURANTE C GALLONIO/FRON(TO)NE Q MARCIO TURBONE LEG/A [U] GUSTI PR PR PER FINES CIVITATIS SERD(IC)ENSIUM REGIONE DYPTENS(IUM)/PRAESIDIÁ N IIII BURGI N XII PHRURI/N CIX. It was found at Bâlgarski isvov, and is datable to 152.

26. K. Bittel & A. M. Schneider, Arch. Anz 56 (1941), 278-80, found at Kirkilise, this inscription was set up under the same governor as that quoted in note 25. ἡ πόλις Βιζυηνῶν κατεσκεύασε τοὺς πύργους διὰ ἐπιμέλητων Φίρμου Αὐλουπορεως κὰι Αὐλουκένθου Δουσακένθου καὶ Ῥαζδου Υακίνθου.

27. See Mihailov, op. cit. (n. 24). The inscriptions are of the same form as that quoted in note 25, but more fragmentary and do not give any numbers of fortifications built. One is datable to 152, and comes from Nersa, in Provadija, recording burgi built round Marcianopolis, and the other, datable to 155, records defensive preparations in the region of col. Flav. Deult(ensium).

28. CIL XIII. 6509 - 'in the style of writing of the late second century'.

29. CIL III. 3385 IMP CAES M [AUR COMMODUS AN]/TONINUS PIUS SAR[MAT]GERM / BRIT PONT MAX TRIB POT X [IMP VII]/COS III P P RIPAM OMNES BU[RGIS]/A SOLO EXTRUCTIS ITEM PRAES [I]/ DIS PER LOCA OPPORTUNA AD/CLANDESTINOS LATRUNCULO/RUM TRANSITUS OPPOSITIS/MUNIVIT PER [..... See also CIL III. 10312 and 10313.

30. CIL VIII. 3, and 2494-5.

31. CIL VIII. 22629. This burgus had evidently been built before 183-5 which is the date of this inscription, since it records that it was rebuilt because the old fortifications had fallen down through decay.

32. IGBR III. 2 no. 1690 (pp. 103-120). See also G. J. Murphy, The Reign of the Emperor Septimius Severus (Philadelphia 1945), 51-4.

33. Murphy, op. cit. (n. 32), 54.

34. Text of the Anonymus Byzantinus is published in H. Köchly and W. Rüstow Griechische Kriegschriftsteller, 2, 2 (Leipzig, 1853), 41-209. See also K. Krumbacher, Handbuch der klassischen Altertumswissenschaft, 9.1. (Byzantinische Literaturgeschichte 2nd ed. Munich 1897) 635-7, and M. Jähns, Geschichte der Kriegswissenschaft, 1 Abt., 143ff. The passage from the Anonymus here quoted is 228b, 4-6. (Köchly and Rüstow give Greek text and German translation).

35. CTh XII. 19. 2. See also M. T. Rostovzeff, JRS 8 (1918), 27ff.

36. CIL VIII. 2495.

37. CIL III. 3653 (dated to 371). On the barbarian dependance on trade, and the disastrous consequences for them if trade with the Roman empire was cut off, see Amm. xxvii, 5.7.

38. See O. Seeck, s.v. Burgus, in PW (-K) III, 1066-7, and M. Labrousse, 'Les burgarii et le Cursus Publicus', Mel. d'Arch. et d'Hist. de l'Ecole Francaise de Rome 56 (1939) 151ff. For provision of mutationes for the imperial post in North Italy in 362-5, see CIL V. 8987 and CTh VIII. 5.2.

39. J. Garbsch, Der spätrömische Donau-Iller-Rheinlimes (Stuttgart, 1970); and also 'Die Burgi von Meckatz und Untersaal und die Valentinianische Grenzbefestigung zwischen Basel und Passau', B. Vbl 32 (1967), 51ff.

40. K. Stehlin, Die spätrömische Wachttürme am Rhein von Basel bis zum Bodensee (Basel, 1957).

41. References are rather scattered: see S. Soproni, 'Der spätrömische Limes zwischen Višegrád und Estergom', Limes Konferenz, Nitra (Bratislava, 1959) 131ff. J. Szilágyi, Aquincum (Budapest, 1956), 24-6, and the references cited in S. Soproni, 'Burgus Bau-Inschrift von Jahre 372 am Pannonischen Limes' Studien zur Militärgrenzen Roms (Köln Graz, 1967), 138ff., and T. Nagy, 'The frontier of Pannonia as reflected by recent research' in ed. S. Applebaum Roman Frontier Studies 1967 (Tel Aviv, 1967), 145ff.

42. CIL XIII. 11573 Summa Rapida at Koblenz, Switzerland; CIL XIII, 11538, (371 A.D.?); ILS 762, (?367): CIL III 10596 (370) and 3653 (371).

43. Amm. xxviii. 2,1 and similar passages xxx. 7.6 and xxviii. 2.2.

44. Eumenius Pan. Lat. ix (iv) 18. 4; Malalas, xii. 308; Proc. Aed. ii. 8. 7. recording the construction by Diocletian of three forts on the eastern limes. Zos. ii. 34, gives a brief résumé of Diocletian's overall policy.

45. H. Hinz and I. Hauberg, Beiträge zur Archäologie des römischen Rheinlands; Rheinische Ausgrabungen 3 (1968), 167ff. H. Schönberger, JRS 59 (1969) Map C. no. 6.

46. Ibid., no. 30. Ant. Class. 17 (1948) 119ff.; NKNOB, 6th Series, 10 (1957), 133ff.

47. ibid., no. 4. OMROL 14 (1933), 11ff.

48. ibid., no. 29. J. H. Holwerda, Nederlands Vroegste Geschiedenis (Amsterdam, van Looy 1925), 219ff.

49. ibid., no. 59. Pfalz. Mus. 36 (1919), 21ff. F. Sprater, Der römische Eisenberg (Speyer, 1952) and Die Pfalz unter den Römern, 1 (1929), 55ff.

50. J. -J. Hatt, Cahiers Alsaciennes d'Histoire et d'Archéologie 132 (1952) 63ff.

51. A. Barb, Jh OAI 37 (1948) Col. 263ff.

52. F. Winkelmann, *Germania* 1 (1917),45; 2 (1918), 54.

53. A. Grenier, *Manuel d'archéologie prehistorique Celtique er gallo-romaine* (Paris, 1924- ) vol. 5,447 and 449.

54. *V. Severini* 4.7. The building at Heiligenstadt is described by A. Neumann, *BJb* 162 (1962), 480ff.

55. See, for example, Macmullen, *op.cit.* (n.8), fig.D, after p.118.

56. *CIL* VIII. 22774. See also similar inscriptions VIII.9725, 19328 and 21531; *CIL* VIII. 8209. On *burgi* in Syria, see Macmullen, *op.cit.* (n.8), 142, n.82.

57. For example the inscription *CIL* XII. 1524 at Sisteron.

58. Perhaps the villa at Froitzheim, which appears to have been surrounded by *burgi*, is one such example, *Beiträge zur Archäologie der römischen Rheinlands; Rheinische Ausgrabungen 3* (1968) 1ff.

## THE INVENTIONS

by Mark Hassall

The object of this paper is to re-open two questions posed by Professor Thompson about the inventions:[1] 'Could the machines have been made and used? If they could, what would have been their value?' To anticipate my general conclusion, I believe that in every case they could actually have been constructed and that as far as the two types of *ballista* at least are concerned, the Anonymus is describing actual existing machines and that a phrase used of the *ballista fulminalis*, that is 'has been found by experience (*usu*) to surpass all others in range and power, (XVIII.1) implies as much. For Thompson, on the other hand 'All attempts to explain these machines (the *ballistae*) should start from the assumption that in practice they would *not* have worked at all' [2], the Anonymus had boldly rejected the concept of torsion artillery, without having clearly thought out anything to put in its place. What I have done throughout this paper is to take exactly the opposite view - to assume that even the most bizarre of the Anonymus' 'inventions' are practicable even if, as in the case of the scythed chariots, they were never actually made. Such an approach may be over-optimistic but it is worth while taking, if only to see where it gets us.

I wish first to discuss the Anonymus' scythed chariots (DRB XII, XIII, XIV) of which there are three versions. The 'Mark I' (Part 2, Pls. X, XI) is drawn by two horses upon which are mounted two riders; the 'Mark II' (*ibid.* Pls. XII, XIII) by one horse with a single rider and the 'Mark III' by two horses, only one of which has a rider (*ibid.* Pls. XIV, XV). All have scythes that are hinged where they are attached to the axle and which can be raised or lowered by ropes passing through rings that are attached to the horses' flanks. Professor Thompson points out[3] that by mounting the driver(s) on the horses greater control is exercised over the machine (and we may add the task of the driver is rendered very much easier), and that the axles will be of the fixed type (the elevating rope would otherwise be rapidly wound round the scythe blade). One point that he does not touch upon is how the horses were attached to the chariots. This is not just a question of idle speculation since I think that by defining the problems and suggesting solutions it may be possible to fill the *lacuna* recognized by Professor Thompson in the Anonymus' description of the Mark III machine. The only (DRB XIV) scholar who has given any thought to the question of traction is G(odescalcus) Stewechius in his commentary on Vegetius published in 1595. In his rendering of the Mark I he adds traces attached to a swingletree.[4] This is unfortunately an anachornism as are the stirrups used by the rider. In fact, as far as the two-horse versions (Marks I and III) are concerned there is no problem. Both would have been provided with normal chariot pole and there would have been a yoke (not shown in the illustration in the MSS.) As long as the horses were sufficiently widely spaced, the inner leg of both riders would not have been crushed against the chariot pole. There is incidentally, a little evidence both literary and archaeological for riders mounted on chariot horses in Roman times. Thus Suetonius[5] says how Tiberius "At the age of nine delivered a eulogy of his dead father from the rostra. Then just as he was arriving at puberty, he accompanied the chariot of Augustus in his triumph after Actium, riding the left trace-horse while Marcellus, son of Octavia, rode the one on the right." An engraved bronze vessel (cista) illustrated in Daremberg et Saglio[6] shows just such a scene.

The Mark II (currodrepanus singularis, Part 2, Pls. XII, XIII) however presents problems. The only way that the single horse can have drawn this vehicle is by being harnessed between shafts. Carts and other vehicles with shafts are known from the Rhenish genre sculpture of the third and fourth centuries but, in the present case, the rider creates obvious though not insuperable difficulties. One solution is for the shafts to terminate short of the rider's legs[7]. The use of shafts though not explicitly mentioned by the Anonymus in his description of the Mark II seems to be definitely implied in his description (XIV.1) of the Mark III which he contrasts with it, for in the latter case the chariot (or as we may suppose, the chariot team) is described as being "less embraced". If this is correct it allows us to restore the sense of the lacuna recognized by Thompson in the description of the Mark III in this section[8].

The Mark III (currodrepanus clipeatus, Part 2, Pls. XIV, XV), like the Mark I has two horses but, unlike it, only one rider. It is also protected at the rear by shields and is fitted with automatic lashes to urge the horses on. The illustrations in the principal MSS. are of little help in trying to decide how the automatic lashes worked and the device mounted on the chariot more resembles a clothes mangle than anything else. If the version given in Gelenius' printed edition is nearer the archetype (reproduced in Part 2, p. 000) then it is not difficult to imagine a simple drive-system for the lashes, especially if one assumes the use of a chain.[9] (Figs. 4 and 5).

All the horses that draw the scythed chariots have been provided with scale horse armour. The representation of this, as is shown by finds of armour of this type at Dura on the Euphrates, is much more accurate than the carvings of Trajan's column.[10]

The mounted warrior driving the Mark III chariot is distinguished from the other chariot drivers by the wings that spring from behind his shoulders. From the tips of these wings hang chains and on the end of at least one of these is a curious object which seems to the present writer to resemble the bronze horse pendants found on Roman military sites of the first and second centuries[11]. But what of the wings themselves? Their presence at first seems inexplicable but one possible explanation is that the Anonymus had read and misunderstood a Greek military manual such as Xenophon's on Horsemanship where πτέρυγες (litarally "wings" but regularly used of the leather straps attached to the lower half of the Hellenistic and Roman lorica) are described in one place[12] as being fitted also the shoulder to allow freedom of movement in hurling a javelin. I think it just possible that the illustration of the Currodrepanus Clipeatus in the editio princeps (reproduced in Part 2,

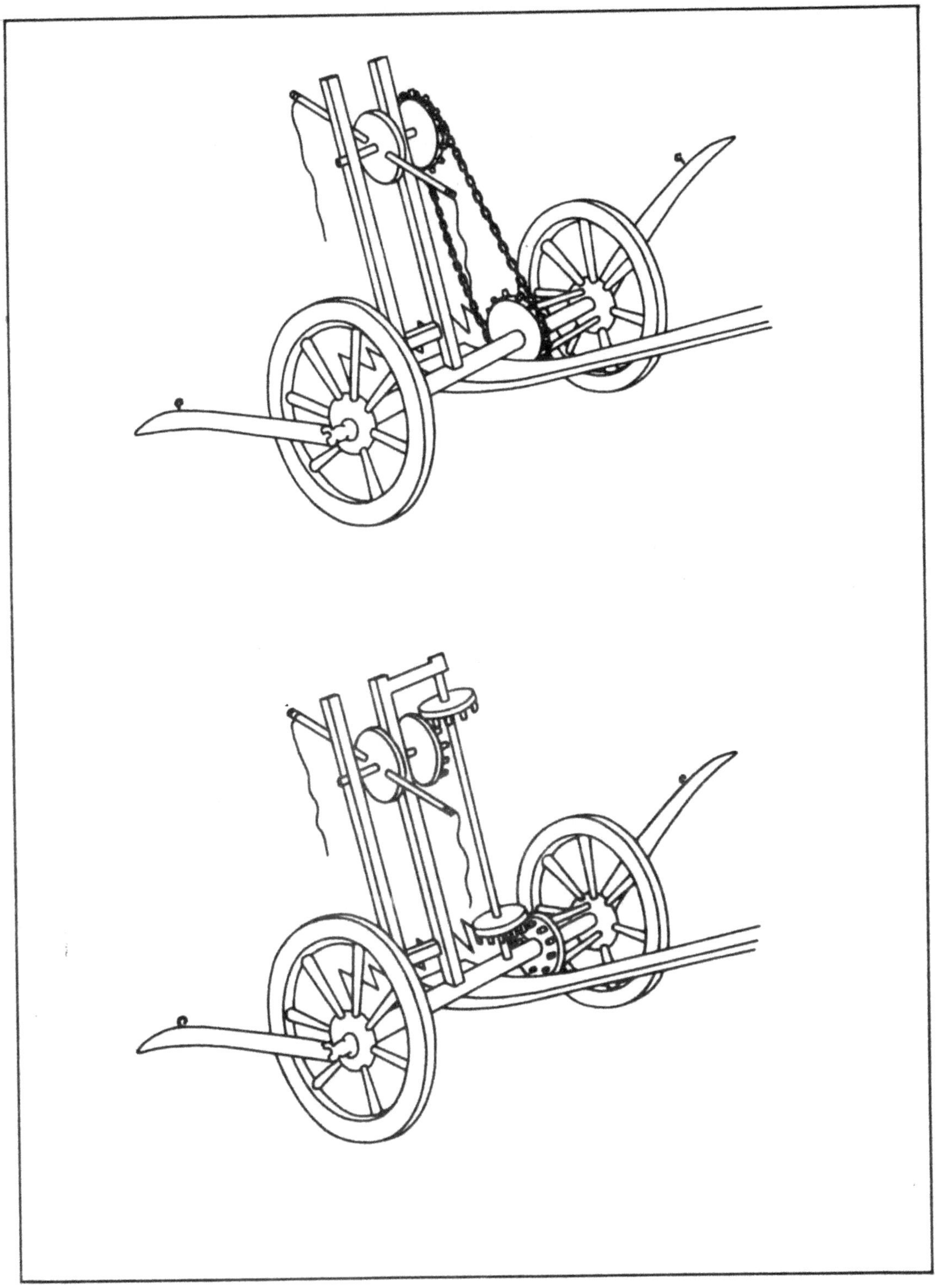

FIG. 4 (above) Drive for lashes of <u>currodrepanus</u> using a chain and toothed wheels

FIG. 5 (below) Alternative drive using toothed wheels
Drawing by Moira Mackenzie

p.13) may have inspired the designers of the armour of the Polish hussars of the seventeenth century, for they too were furnished with wings (Fig. 6).

Turning from the scythed chariots to the ballistae, we find that the Anonymus describes two types, the heavy static ballista fulminalis and the mobile ballista quadrirotis. Of these it is the fulminalis (XVIII, Part 2, Pls. XXII, XXIII) that has provoked more discussion. For Thompson the fulminalis represents a gallant but naive attempt on the part of the Anonymus to break away from the principles of conventional Roman ballistae and onagri with their springs of twisted sinew, but he does not believe that our author had anything better to offer in their place. Marsden[14] thought that both the ballistae of the Anonymus were torsion engines while R. P. Oliver[15] argued that the iron arch, which in both machines rises above the 'barrel', was actually a steel spring which supplied the necessary propulsive force. The last theory has been sufficiently refuted by Marsden[16]. Reinach's commentary[17] on the fulminalis is somewhat obscure and he does not clearly explain how he believed the motive force was produced. However, it is virtually certain that he is correct in one essential. Reinach pointed out that there are two pairs of pulley blocks at the rear of the machine and that one pair will have been used to pull back the slider (= the sliding barrel upon which the arrow to be fired rested as in conventional ballistae). When the windlass set behind the pulleys was turned in the opposite direction this second pair will have pulled the slider back to its original position at the front of the machine. From this, incidentally, it follows that in the illustrations the length of the rear part of the machine has been drastically reduced, for the distance between the pulley blocks of each pair must be equivalent to the distance that the slider has to be pulled back, i. e. approximately the length of the slider.

Both Reinach and Thompson translate the cochleae attached to the front of both machines as screws used to elevate them, while Marsden interprets them as univsrsal joints, with some form of which it seems reasonable to assume they were provided. The screw threads, however, are clearly shown on some of the illustrations of the fulminalis and the interpretation as screws here seems on the whole more likely. This implies the existence of a pivot at the centre of gravity of both ballistae. One of the objections that Marsden raises is that the provision of screws creates difficulties when one comes to traverse the machines. In fact this is not really a problem if a universal joint was provided at the centre of gravity.

Marsden[18] is surely correct in implying that the Anonymus and his artist here confused the simple hook connecting the end of the winch-rope to the rear of the slider and the trigger-mechanism. The function of these two parts can hardly have been combined in the same apparatus.

At the side of the fulminalis is what looks like a large cogwheel. This can however equally well and perhaps more plausibly be interpreted as a ratchet-wheel perhaps connected to a small windlass omitted from the illustrations. The possible function of this subsidiary windlass and ratchet will be discussed below.

The biggest puzzle about the fulminalis is how the missile was actually propelled from the machine and Thompson says that this is not described in

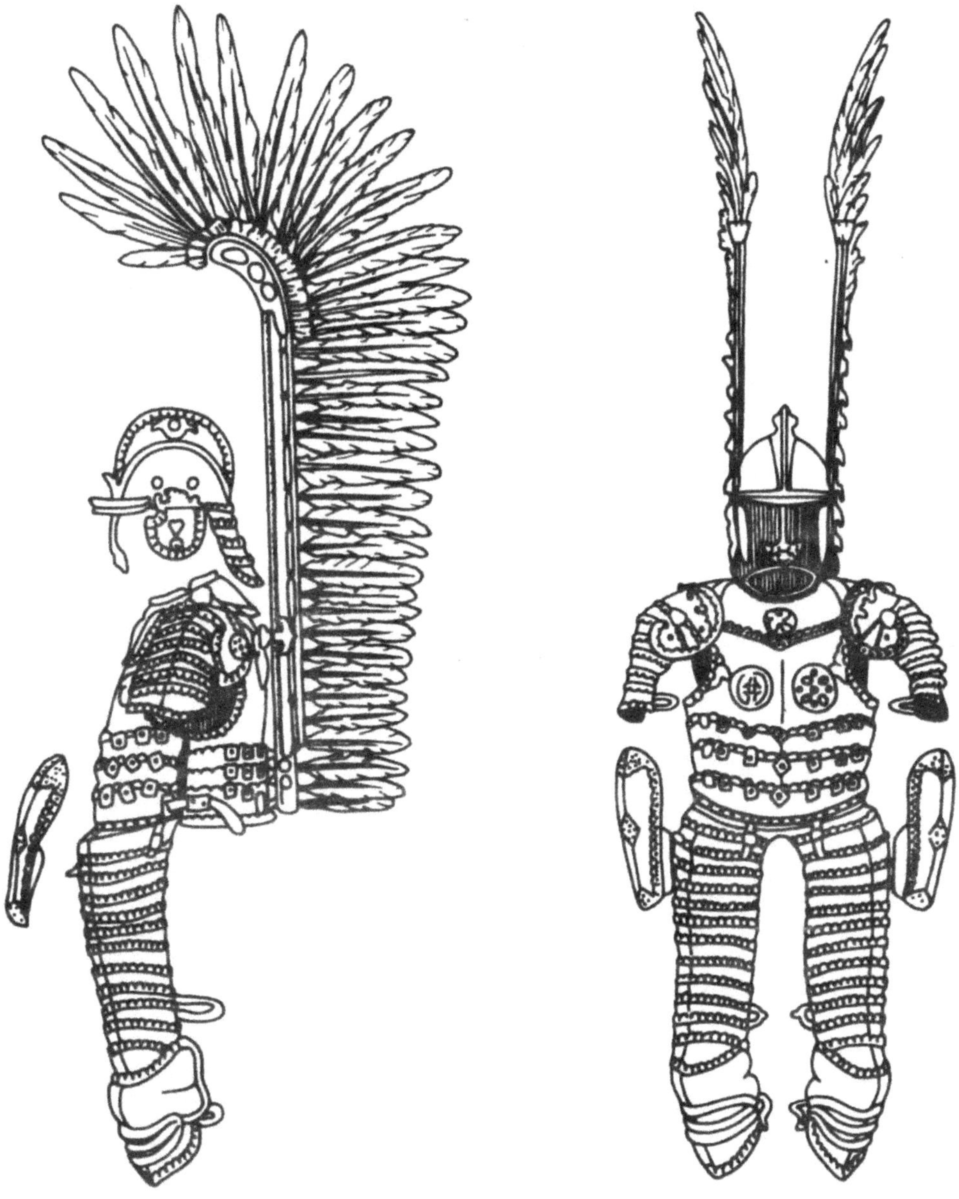

FIG. 6 Armour for Polish Hussar C17th.
Drawing by Moira Mackenzie

the text. This may not be correct, for it could be that the working principles of the ballista can be deduced from a literal reading of the phrase (XVIII. 1) validus nervi funis ferreo unco tractus eandem sagittam magnis viribus in hostem dimissus impellit, 'and a strong rope of sinew is pulled back with an iron hook and fires this arrow with enormous power into the enemy when it is released', and a similar literal interpretation of the illustration. In other words, the ballista could have worked exactly on the same principle as the simple schoolboy's catapult, in which a length of rubber stores the energy when pulled back, and itself provides the motive force when released. Crucial to this explanation is the interpretation of the word nervus. In a transferred sense the word is used of the string of a bow but the basic meaning, like that of the etymologically connected Greek word νεῦρον, is sinew or tendon from which bow strings were often made. Sinew or tendon would be unsuitable for the purpose auggested here, but the ligaments of certain animals would make viable alternatives to rubber and perhaps could have been described as nervi. The ligamentum nuchae of ungulates (which forms one of the elements in the construction of composite bows) runs from the head along the top of the neck to the animal's back. Its function is to counteract the weight of the head of the animal, and relieve the neck muscles. It is made of almost pure elastin, an elastic protein found in vertebrates which "can be stretched a long way like rubber, and snaps back to its initial length when it is released"[19]. Through the kindness of Dr. Pat Hodson and Brigadier Harry Merlin of the Royal Veterinary College I obtained the ligamentum nuchae of a horse and initial experiment has been promising.[20] The function of the subsidiary windlass with its ratchet set on the side of the machine will, on this interpretation, have served to stretch the ligament into position initially and/or tighten it periodically if creep becomes a problem.

The last problem with the fulminalis concerns the main windlass that draws back the slider. The two men operating it appear to be confined in some sort of cage which Thompson claimed was not part of the windlass. This matter is not quite so certain. If, as Stewechius thought[21], the men were enclosed in tread wheels, then this fact taken in conjunction with the pair of multiple pulley blocks used to pull back the slider, implies the application of considerable force with singificant implications for potential power and range of the machine. Finally, it is worth noting that it appears to be extra ordinarily similar to the description of the ballista given by Ammianus Marcellinus, though this in itself presents difficulties of interpretation[22].

Figs. 7 and 8 are based on photographs of a working model of the ballista fulminalis, built according to the ideas outlined above (in both Figs. one of the pulley blocks is almost completely obscured by the back of the slider). The box-like body of the machine is mounted on a universal joint at its centre of gravity, the joint being itself fixed to a cruciform wooden base. The construction of such a model of course only shows that the ballista could have functioned in the way proposed, not that it must have done so.[23] Other interpretations may be suggested by a study of the recent exciting finds of ballista parts made at Hatra in Iraq, sacked by the Persians in 245[24] and at the fourth century sites of Gornea and Orsova in Roumania.[25] The Roumanian finds are particularly relevant to the de Rebus Bellicis since they come from two fortlets of the type recommended by the Anonymus (XX), fortlets, moreover,

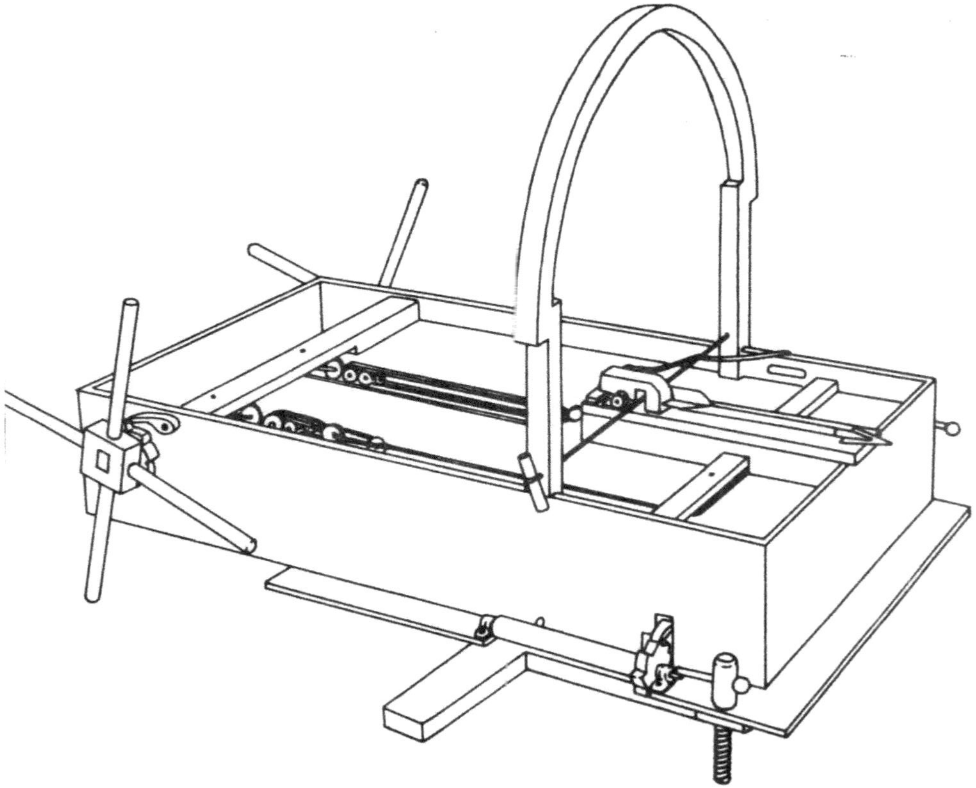
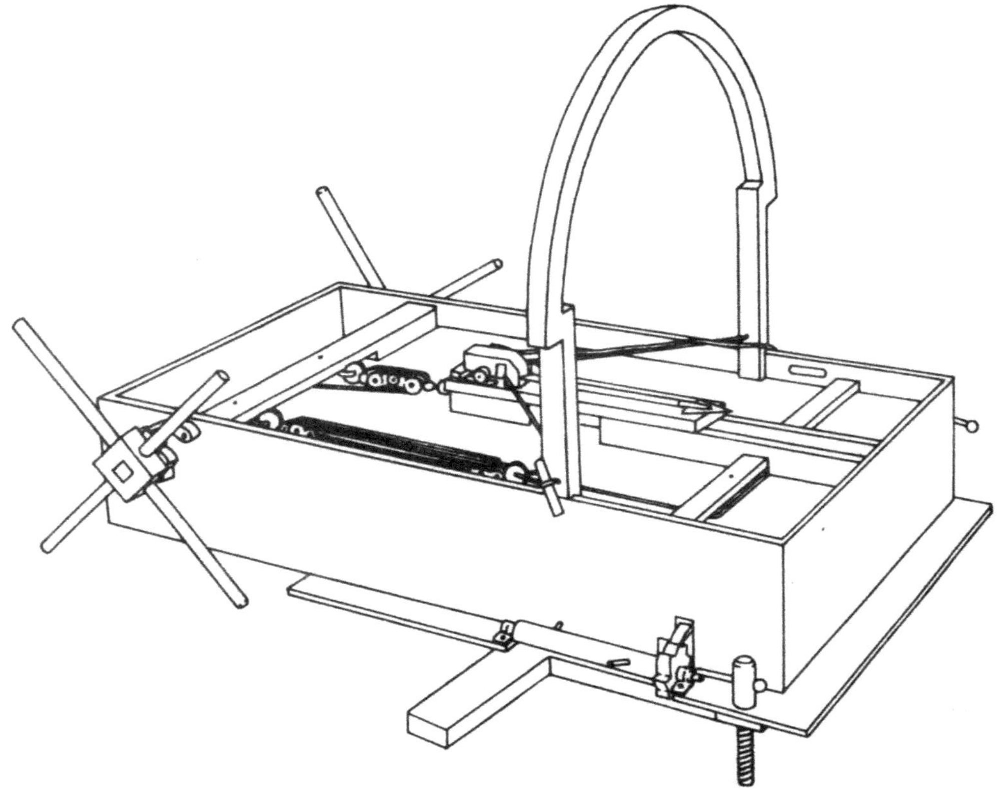

FIGS. 7 and 8. Model of *ballista fulminalis*. Left (FIG. 7) in loading position.
Right (FIG. 8) in firing position.
Drawings by Moira Mackenzie

that were built overlooking the Danube, which is precisely where the *ballista fulminalis* was thought of as operating (XVIII. 5).

The *ballista quadrirotis* (VII) also presents difficulties which are not resolved by the MS illustrations (Part 2, Pls. V, VI). As with the *ballista fulminalis*, there are no obvious vertically mounted torsion springs into which bow arms are inserted in the manner of conventional *ballistae*. How then did the action of turning the windlass at the rear of the machine provide and store the energy which, when released, propelled the missile? One feature present in all versions of the illustration looks in some of them, especially C (Part 2, Pl. Va), like a forward pointing bow arm, but the bent arm of a re-curve bow and not the straight, rigid arm that one would expect. If this interpretation is correct, it might explain a difficult phrase in the description given by the Anonymus (VII. 6) where the missible is said to be discharged from the *ballista, non, ut aliae funibus sed radiis intorta*,[26] where *funes* would then refer to torsion springs and *radii* to re-curve bow arms. However, the bow arm interpretation is much less convincing if we turn to the other versions of this feature, especially that of the second set of illustration in the Munich MS (Part 2, Pl. VIb). Here it looks like a length of rope attached to the front of the slider and passing back from it to the body of the machine and is best thought of as the rope used to pull back the slider after the missle has been discharged, a device with which the *ballista fulminalis* was also provided. The basic question, therefore, of how the *ballista quadrirotis* actually worked remains unanswered.

There are one or two other points to notice. The extension of the flighted end of the arrow back to a point just in front of the windlass, in the illustration in C, is not confirmed by the illustrations in the other MSS, and I take to be a mistake on the part of the copyist. The screw device at the front of the *quadrirotis*, mentioned in the description given in the text, is similar to the *cochleae* on the *fulminalis*, but though recognizable in the illustrations has also suffered at the hands of the copyists. The presence of this screw once more implies the existence of a pivot at or near the centre of gravity of the upper part of the machine. Finally, the versions of the illustrations in P and in Gelenius' edition (reproduced in Part 2, p. 8) show two rings attached to the visible side of the *ballista*, presumably balanced by a pair on the other side. These could possibly be the *quattuor foramina*, acting as handles for the gun crew by which the top of the *ballista* was rotated about a vertical pivot, for by use of these *foramina* we are told the machine was 'turned round and aimed' (VII. 3).

The one machine upon which the repuration of the Anonymus as a serious inventor rests is the *liburna* driven by paddle wheels (XVII). Unlike the inventions discussed so far this really presents no problems. It is worth, however, turning to the illustration (Part 2, Pls. XX, XXI) that supplements the texts for some points deserve comment. The sails themselves have been reefed by means of brailing ropes and, to allow us to see them properly, have been turned by the artist through 90°. This is a convention that we shall meet with again when looking at the *tichodifrus*. In general the vessel bears a remarkable resemblance to modern reconstructions of Roman biremes, if one thinks away the paddle wheels. The presence of an "outrigger" for the upper

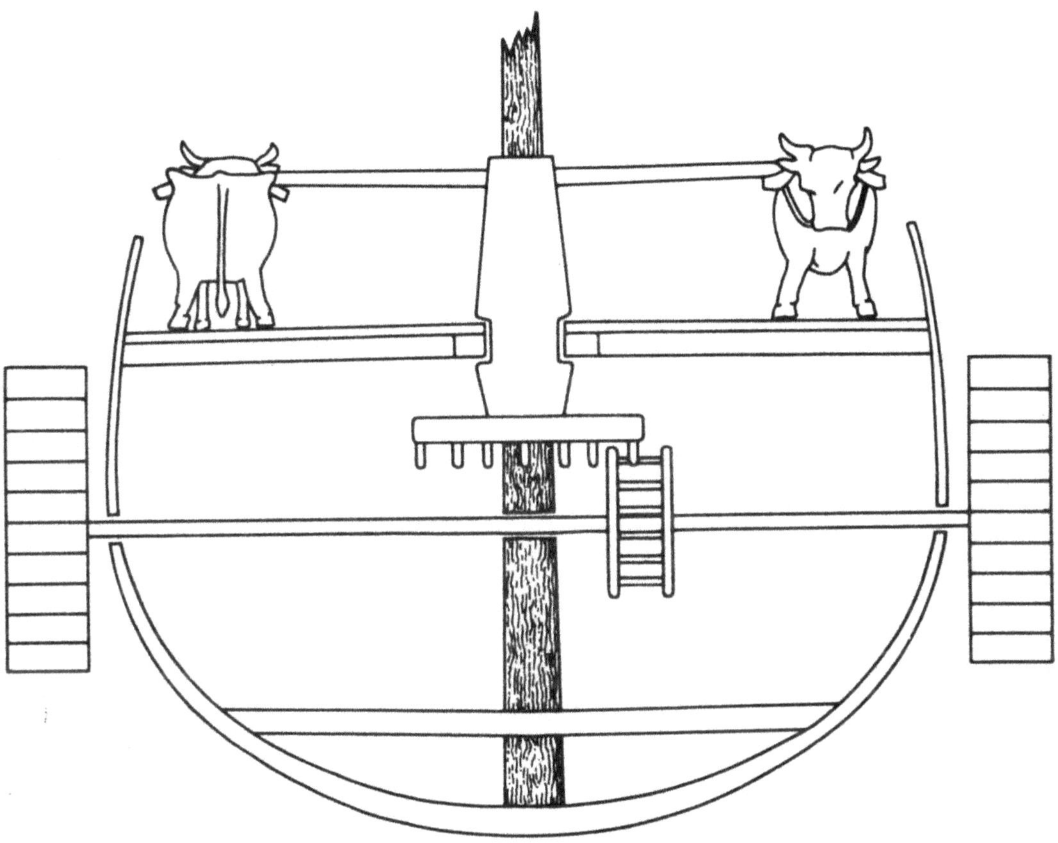

FIG. 9 The liburna, gearing for central capstan
Drawing by Moira Mackenzie

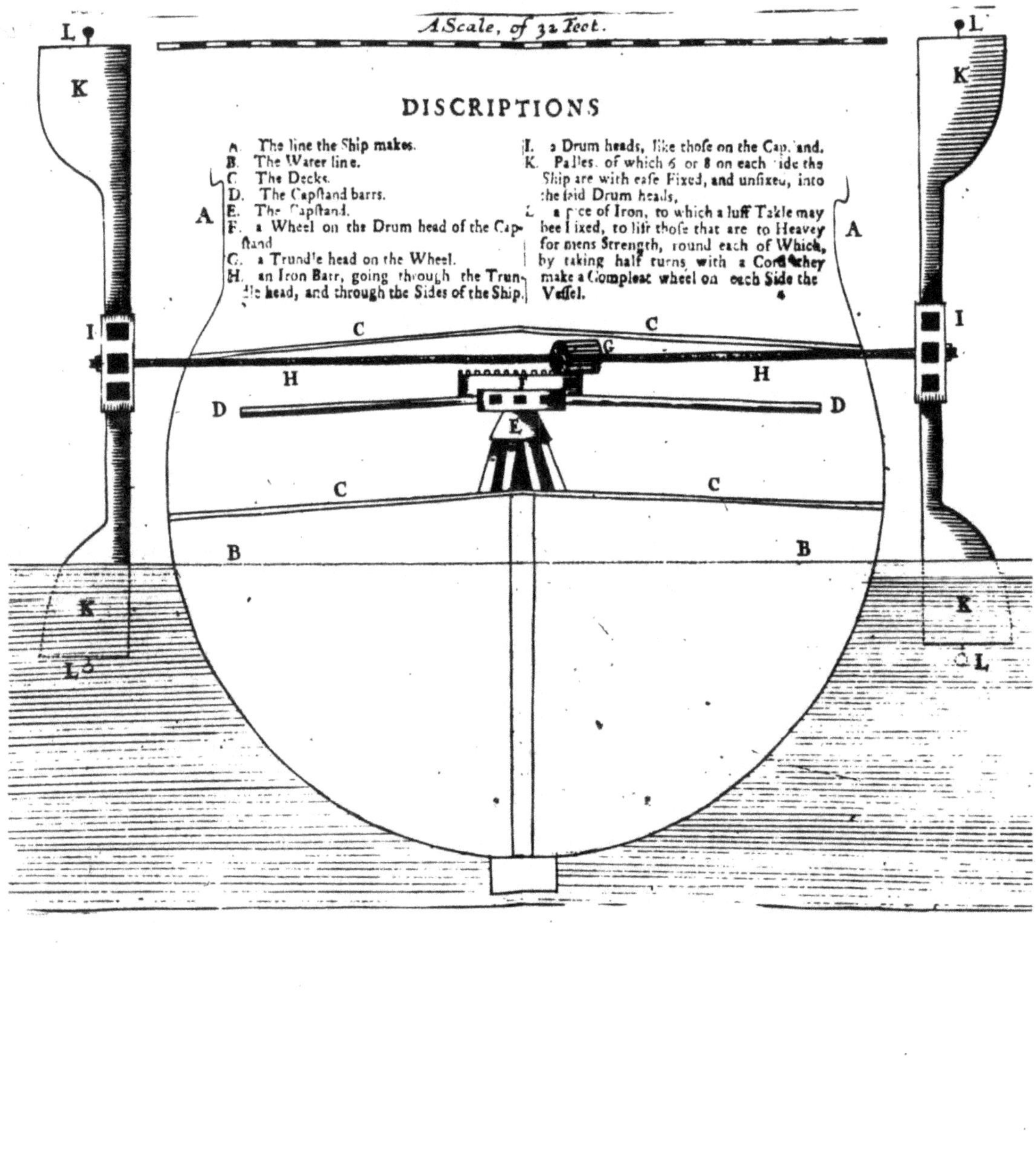


FIG. 10 Thomas Savery's geared paddle-boat
(from Navigation Improv'd)

set of thwarts is an interesting detail,[27] as is the small secondary ram attached to the prow of the vessel. Just such a bronze subsidiary ram has recently been deposited on loan at the Fitzwilliam Museum.[28]

The gearing system used for transmitting the rotary motion of the capstans driven by the oxen to the paddle-wheels is simple and, as Thompson points out, is identical to that of the Vitruvian watermill, and, we might add, geared water lifting wheels turned by oxen.[29] Drive systems of this sort are attested archaeologically.[30] The only difficulty in reconstruction is caused by the positioning of the central capstan which is placed awkwardly to one side of the central mast which would prevent the oxen from turning it. In fact, if the Anonymus had so desired, the capstan could have formed a sort of sleeve round the mast.[31] (Fig.9)

How effective would the vessel have been in practice? The conclusion of Prof. Gordon of Reading University[32] is that "simple calculation .... shows that the speeds attainable would be too low to be of any military use." Be this as it may, paddle-driven boats caught the imagination of inventors and engineers of the Middle Ages, though at first no practical experiments were made. Reinach[33] ascribed the first serious attempt at making an actual boat operated by paddles to the French engineer Duquet in 1699. In fact Duquet was anticipated by at least three and possibly four inventors. One of the earliest is also the most enigmatic: the paddle boat powered by steam supposedly built by the Spaniard Blasco de Garay in 1543. The genuineness of this claim made first in the nineteenth century has been doubted.[34] But of other early paddle boats we have contemporary or near contemporary circumstantial accounts. For example, there was the "Castle of Delft" built by the Dutch admiral Boisot (died 1576), a double hulled vessel propelled by a centrally placed paddle wheel geared to a windlass turned by twelve men.[35]

Of particular interest is the boat built on the orders of Prince Rupert of the Rhine, First Lord of the Admiralty (1673-1679). This is referred to by two pioneers of the steam engine, Thomas Savery and Denis Papin,[36] both of whom were extremely impressed by the vessel's performance and both of whom later built (man-powered) paddle boats of their own. (Fig. 10). From their accounts we learn that the Prince's vessel drew 4½ feet of water and propelled by four, six or eight horses; that in a contest on the Thames it easily outstripped the Royal Barge rowed by sixteen men and that it was in use in 1682 as tug for even the largest vessel at Chatham (where the winding course of the river made it difficult to manoeuvre a ship under sail).

In the late eighteenth and early nineteenth centuries there were at least three areas where horse driven paddle boats were in use. Across the Atlantic, a river boat built by a Dutchman was in use on the Ohio[37] while several double hulled vessels driven by horse-powered paddles were in use in 1814 as ferries at New York.[38] Finally, there was the Great Yarmouth Horse Packet cited by Thompson[39]. This vessel was about 60 feet long and 18 feet in the beam. It was propelled by four horses turning a capstan and went at about six miles an hour.

Of these paddle boats, that built by Prince Rupert was, as far as we know, the first to have been propelled by animal power. It seems not impossible that the Prince was inspired by the de Rebus Bellicis, perhaps known to him

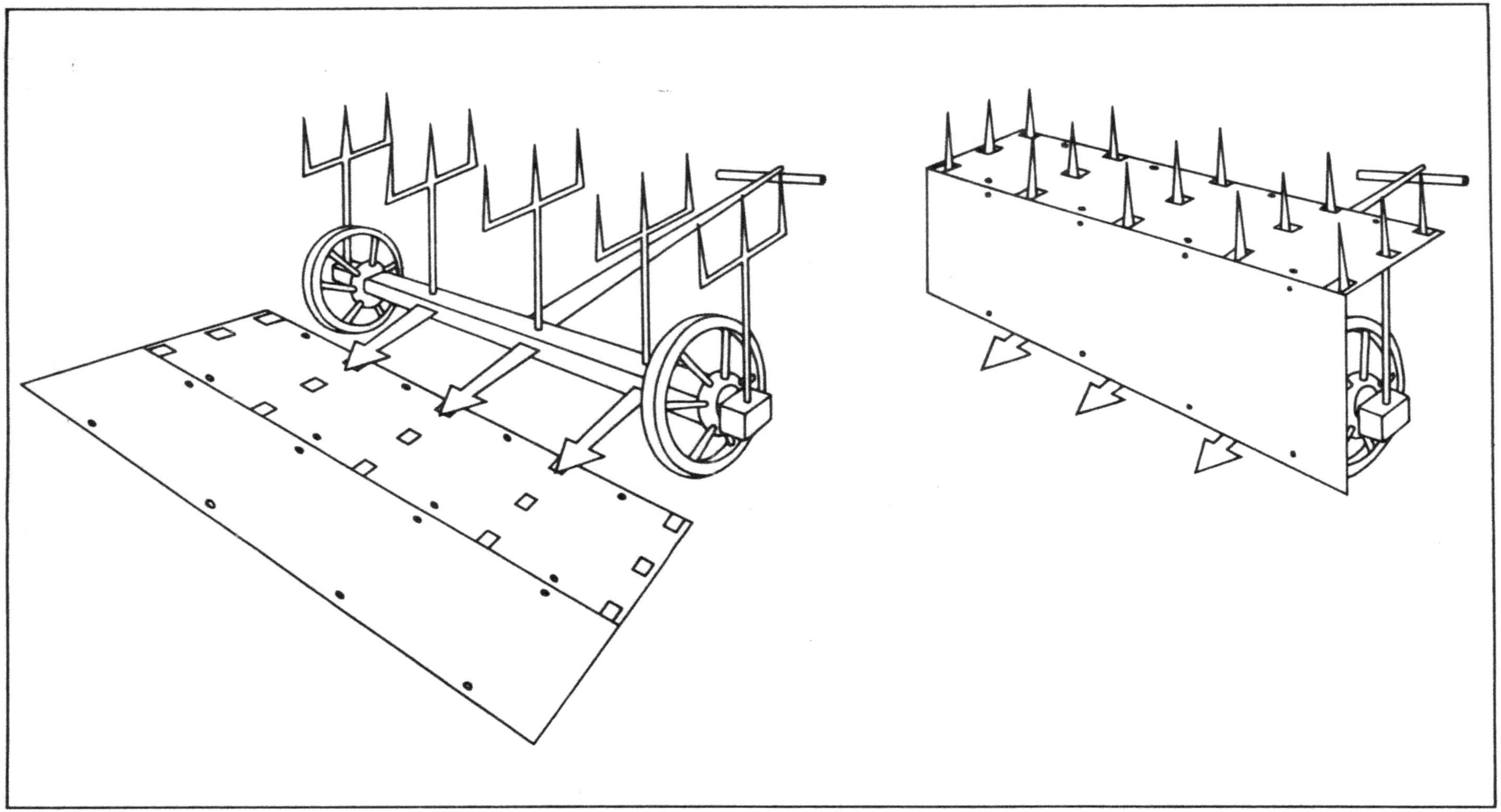

FIGS. 11 and 12 The _tichodifrus_ with _crates_ removed and in position

through Gelenius' printed edition, or even the Codex Spirensis itself, for he was related to Prince Otto Heinrich who acquired the book from the Cathedral library at Speyer. Rupert's boat was powered by horses, as were the North American and Great Yarmouth paddle-boats and not oxen like the liburna of the Anonymus. This is hardly surprising given the great improvement in the breed of draught horses that took place during the Middle Ages when a horse was required that could carry the weight of a knight in full armour. In the nineteenth century, engineers reckoned that a good draught horse and an ox could both exert a force of 120 lbs. but the horse moves faster and is therefore more effective as a prime mover.[40] However, in antiquity before these improvements in the breed of horses took place, the ox whose harnessing arrangements were also more effective, was the principal draught animal and therefore the natural choice of the Anonymus for his paddle boat.

There is much less to say about the two final 'inventions'. The tichodifrus literally means 'wall chariot' but the Anonymus gives a false etymology for he says (VIII.1) that 'its function corresponds with the meaning of that word: that is to say that a wall can easily be climbed by means of it' (quod per hunc facilior in murum paretur ascensus). Though he knew the meaning of τεῖχος even if he could not always spell it, he clearly did not know the meaning of δίφρος, a chariot, so called because it carries two persons. Possibly he was misled by the similarity of the word to the Latin defrusto = divide in pieces, dismember, a word used by Ammianus[47]. It is interesting to note that in XIX.5 he describes the tichodifrus as 'carried along by the agency of two men', which looks as if it could derive from a marginal note or actual glossary explaining the true derivation of the word.

The interpretation of the illustration given in the MSS (Part 2, Pls. VII and VIII) is difficult. In the text we read that (the sides of) the vehicle do not stand high off the ground so as to conceal the vestigia, that is the feet of those walking inside (VIII.2). This means that the vehicle lacked a base, yet was protected by some form of superstructure. However, as Thompson points out,[42] the illustration seems to show the chassis alone. What the superstructure was composed of the Anonymus does tell us—two hurdles placed on top and fastened (or studded) with nails (duabus superimpositis cratibus fixoriisque confixis). Now set under the lanceae which project from the front of the machine is a flat rectangular object divided by a dark longitudinal line, upon which are four rows of fixoria. The identification of these last is assured by the illustration of the clipeocentrus where the radially arranged fixoria minuta described in the text are clearly shown. It seems possible that this object represents the two crates, perhaps hinged together, resting on the ground before being 'superimposed'. If this is correct, they could have been supported by the curious tridents with their odd button-shaped terminals with which the machine is furnished. These may have been turned through 90° by the artist so that their nature could be seen more clearly, as the wheel on the left side of the machine has been and as was the case with the sail of the liburna. The terminals, or perhaps rather the cross-bar of the tridents would support one of the crates placed horizontally over them while the second would hang down in front vertically with its bottom edge resting on the rear part of the lanceae. (Figs. 11 and 12).

How was the tichodifrus propelled forward? If the phrase 'borne by two men' is to be taken at its face value the men in question would manhandle the

FIG. 13  Portable Chinese raft made of inflated goatskins

*tichodifrus* forward by holding onto the pole with one hand and the cross-piece with the other. However, if, as I have suggested above, this phrase only derives from a learned note on the word δίφρος found in a source, then it is possible that the machine should really have been propelled by two horses, pushing forward in the manner of the horse who drives the Gallic reaping machine or *vallus* and harnessed in a similar way[43] (Fig. 1).

Finally, the inflatable bridge, (*ascogefyrus*), described in XVI. There is little to add to Thompson's Commentary except to draw attention to the version of the illustration given by Stewechius.[44] This shows buckles fastening leather straps holding the constituent inflated leather bladders together, which seems a reasonable interpretation of the illustrations in the MSS (Part 2, Pls. XVIII, XIX) although it does not tally with the description given by the Anonymus which seems to refer to hooks and eyes (XVI. 2: *de super autem parte circulis exstantibus, ex altera immittuntur uncini*). The article by Hornell cited by Thompson[45] admirably demonstrates the practicability of the *ascogefyrus* as well as the actual use made of comparable devices from antiquity to relatively modern times. Henry Hodges has also drawn my attention to a picture of a relatively modern example of a raft supported by inflated bladders which bears an altogether uncanny resemblance to the Anonymus' inflatable pontoon (Fig. 13)[46]. Since this is carried by one man, some idea of the sort of width of river that the Anonymus envisaged spanning with his bridge can be gauged from his statement in the perface to the *de Rebus Bellicis,* that it could be carried dismantled by fifty pack-horses (*Praef*. 14).

Institute of Archaeology,
31-34 Gordon Square,
London WC1H 0PY.

## NOTES AND REFERENCES

1. Thompson, 78.

2. *ibid.*, 64. Elsewhere Thompson takes a much more optimistic view particularly of the inflatable bridge and the paddle boat.

3. *ibid.*, 57 and n.1.

4. G. Stewechius, *Commentarius, ad Flavi Vegeti Renati Libros de Re Militari* (Antwerp, 1585), 264 (illistration).

5. Suetonius *Tib*. 6.

6. Daremberg et Saglio, *Dictionnaire des Antiquités Grècques et Romaines*, Vol. 1, part 2(c), fig. 2223.

7. See Le Comte Lefebvre des Noettes, *L'attelage. Le cheval à selle à travers les ages. Contribution à l'histoire de l'esclavage*. (Paris, 1931), fig. 161.

8. See the translation of this chapter in Part 2.

9. Note the use of a chain in the repeating catapult of Dionysias of Alexandria. E.W. Marsden, *Greek and Roman Artillery Technical Treatises*, 147-53, diagram 9, (Oxford, 1971). cf. The marble-cutting saws driven by the water power provided by the river Ruwer near Trier in the fourth century (Ausonius, *Mosella*, 361-364), which, if they were not circular iron discs, could have been wire band-saws. In either case they probably did not have teeth, but worked with an abrasive.

10. For Dura see *Excavations at Dura-Europos*, *Report of Sixth Season* (Yale, 1936), Pl. XXII and for Trajan's Column, C. Cichorius, *Die Reliefs der Traianssäule* (Berlin, 1896 and 1900), cast nos. 92-94.

11. See for example G. Webster, *The Roman Imperial Army* (London, 1969), 152, Fig. 23. There are minor variations in the MSS illustrations (Part 2 Pls. XIV and XV) and those of M* omit them altogether. The versions in M resemble the large tassels.

12. Xenophon, περὶ Ἱππικῆς 12.6. They are of course often represented on the monuments.

13. Thompson, 64, 65.

14. Marsden, *op. cit.* (n. 11), 235.

15. R.P. Oliver, 'A Note on the *De Rebus Bellicis*', *CP*. 50 (1955), 113ff.

16. Marsden, *loc. cit*. For the awareness of spring steel in Hellenistic times see H. Bartlett Wells, 'Knowledge of spring steel in Hellenistic times', *Arms and Armour*, 7 (1971-3), 249-256. Numerous finds of Roman locks in Britain and elsewhere show that spring steel was widely used for this particular application during the Roman period.

17. S. Reinach, 'Un homme d'idées au Bas-Empire', RA ser. 5.16 (1922), 205-65.

18. Marsden, op.cit. (n. 9), 245 n.3.

19. R.McNeill Alexander Animal Mechanics (London, 1968), 81. Young's modulus is about 6 x $10^6$ dyn/$cm^2$. H. Yamada Strength of Biological Materials (Baltimore, 1970), 92, 93, gives further data on the ligamentum nuchae of cattle, including the stress strain curve. The highest value for the ultimate elongation is 160%. He also gives data on the elastic properties of this material after being treated in various ways. Thus tissue stored in 20% solution of acetic acid (= vinegar, the acetum of the Romans) for four days maintains 98% of its elasticity and 90% of its strength.

20. The ligamentum nuchae was kept supple by storing it in a 20% solution of acetic acid.

21. Stewechius op.cit. (n. 4), 321 (with Fig.)

22. Amm. 4. 1-3 see the text translation and illuminating notes given by Marsden op.cit. (n.11), 237-240. I do not, however, agree with him on every point.

23. Compare the remarks by John Coles in his introduction to Archaeology by Experiment (London, 1973), 17.

24. D. Baatz 'The Hatra Ballista' Sumer 33 no. 1 (1977), 141-51.

25. idem 'Teile spätrömischer Geschütze aus Rumänien' Arch. Anz. 3 (1975), 432-4, and 'Teile spätrömischer Ballisten aus Gornea und Orşova (Rumänien)', Saalburg-Jahrbuch 31 (1974),50-72. Baatz has recently published an account in English of both the Hatra and Rumanian finds, see Britannia 9 (1978), 1-17.

26. DRB VII. Marsden op.cit. (n.9), 243 suggests that the radius was a toothed iron bar used in conjunction with a cog wheel for pulling back the slider. Thompson, 114, translates radiis intorta as "propelled ... by a windlass" i.e. he interprets the radii as the handspakes of the windlass.

27. Cf. for example, B. Landstrom, The Ship (London 1961), fig. 95 (outrigger) and contrast A.L. Ben-Eli and R.R. Stieglitz (eds.) The National Maritime Museum, Haifa, (London, 1972), Pl. 38 (no outrigger).

28. R.V. Nicholls, "The Trinity College Collection and other recent loans at the Fitzwilliam Museum", in Archaeological Reports, 1970-71, no. 17, (Hellenic Society and British School at Athens), 85, fig. 14.

29. T. Schiøler Roman and Islamic Water-lifting Wheels (Odense 1973).

30. L.A. Moritz, Grain Mills and Flour in Classical Antiquity (Oxford, 1958), plate 14c, for the working parts of a mill from the Saalburg.

31. This is the solution adopted by my friend Mr. Paul Poulain who has built a model of a liburna, driven by a single capstan and pair of paddle-wheels. A single pair of paddle wheels, incidentally, would not be affected by turbulence in the water as would have been the case with the two rear pairs of paddles of the Anonymus' liburna.

32. Stated in duplicated lecture notes on the History of the Warship.

33. op.cit. (n.17).

34. see H.P. Spratt, The Birth of the Steam Boat (London, 1958), 18ff.

35. Nicholas Witsen, Architectura Navalis et Regimen Nauticum,(Amsterdam, 1960), 38. This book has recently been reissued in a facsimile edition, Amsterdam (no date). Just such a double-hulled paddle vessel was built by Patrick Miller in 1787. His specifications were republished by Bennet Woodcroft in Steam Navigation (1848) and there is a contemporary model in the Science Museum, London.

36. Denis Papin, 'Nova methodus ad vires motrices validissimas levi pretio comparandis', in Acta Eruditorum, 91 (Leipzig, Aug. 1690), 413. Thomas Savery, Navigation Improv'd: or the Art of rowing Ships etc. (London, 1698). For the fate of Papin's own vessel, see R.L. Galloway, The Steam Engine and its Inventors (London, 1881). Savery includes an illustration of the working parts of his projected vessel in which he hoped to interest the Admiralty (see Fig. 10).

37. Francis Baily, Journal of a Tour ... of North America in 1796 and 1797. (London, 1856), 372, and (London and Amsterdam, 1969, ed. Jack D.L. Holmes), 182.

38. Robertson Buchanan, A practical treatise on propelling vessels by steam etc. (Glasgow, 1816), 70.

39. Thompson, 54. I am grateful to Mr. A.A.C. Hedges, F.L.A., Borough Librarian and Curator, the Central Library, Great Yarmouth, Norfolk, for sending me further details of this interesting vessel, the "Courier".

40. W.J.M. Rankine, Useful Rules and Tables (London, 1866), 251. The average draught horse exerts a force of 120 lbs. at a speed of 3.6 feet per second, or nearly $2\frac{1}{2}$ m.p.h. over a long period while the ox does so at only 1.6 m.p.h. (Note that Rankine's values for a horse driving a gin or mill are somewhat reduced, being 100 lbs force at 3 feet per second). It would be interesting to estimate the probable dimensions of the Anonymus' vessel and using data such as the above calculate the theoretical speed. Taking into account the recorded performance of actual horse-powered boats (Prince Rupert's and the Great Yarmouth ferry), it seems that Gordon's estimates on the performance of the liburna are too pessimistic.

41. Note the reading tichodefrus in the caption of the picture in C.

42. Thompson, 67.

43. For the vallus see pp.39ff. above. The general appearance of the tichodifrus will then have been similar to that of the mobile cannon illustrated in the Encyclopaedia Britannica (11th edition) 2. Pl. I.3 opposite p. 688 = Napoleon III's Etudes sur ... l'Artillerie 3, 178, Pl. 18.

44. G. Stewechius op.cit. (n.4) 207 with Fig.

45. J. Hornell "Floats and Buoyed Rafts in Military Operations", Antiquity, 19 (1945), 72-79.

46. A.M. Cable and F.L. French, China, her life and People, (London, 1946), 72.

# THE *PLUMBATAE* FROM WROXETER

by Philip Barker

As is well known to students of the later Romany army, certain infantry units were provided with short hand-thrown javelins known as *martio-barbuli* ('Mars-barbs') or *(hastae) plumbatae* ('leaded' javelins) from the lead weight attached to the shaft to give the missile greater impetus. According to Vegetius,[1] a group of five such missiles was carried in the hollow of the shield and they were thrown at the first charge. The Anonymus describes two versions of this weapon. The first, the *plumbata tribolata* (X) had, as its name implies, iron spikes fixed in the lead weight so that if it failed to inflict damage as a missile, it could still hurt the enemy by acting as a caltrop. The caltrop *(tribulus)* in question is not the large wooden obstacle mentioned in the literary sources cited by Thompson,[2] but the small iron object known from finds at Roman military sites such as Newstead.[3] The second type, the *plumbata mamillata* (XI), lacked the spikes and had a distinctive head 'round in section and tapering to a point.... powerful enough to penetrate very easily the enemy's shields and similar obstacles'. In other words the head was like that of the sharp, armour piercing, *ballista* bolt heads attested archaeologically, which are also heavy, tapering and needle sharp, only normally square rather than round in section.[4] The epithet *mamillata*, 'breasted' presumably refers to the bulbous lead weight attached to the shaft, although this is omitted in the MS illustrations of the weapon (Part 2, Pl. IX).

*Plumbatae* heads have been found at a number of sites in Britain and on the continent in third and fourth century contexts but no site appears to have yielded so many as Viroconium (Wroxeter, Shropshire). Here, excavations conducted on behalf of the Department of the Environment have shown that there was a succession of timber buildings, some of them very large, covering the site of the *basilica* of the public baths which had probably been demolished during the first half of the fourth century. Among the finds from the latest levels[5] there were four *plumbatae* to which number can be added another found by Dr. Graham Webster[6] in his excavations of the baths themselves and one found during the excavations of the nineteenth century.[7] Taken together these finds probably indicate the presence of a Roman army unit stationed here as late as the early fifth century. They also provide the clearest evidence for the sort of weapon of which the Anonymus' two missiles were adaptations.

The Wroxeter *plumbatae* heads (Pl. XI) are approximately 12 cm long. The tip is arrow-like in proportions and barbed. It is joined to the socket by a long, slender stem and the socket is enclosed, and therefore obscured, by an ovoid lump of lead. The lead prevents normal X-ray examinations but, by courtesy of the A.W.R.E., Harwell, three *plumbatae* were submitted to neutron radiography, an as yet little used technique in archaeological practice.

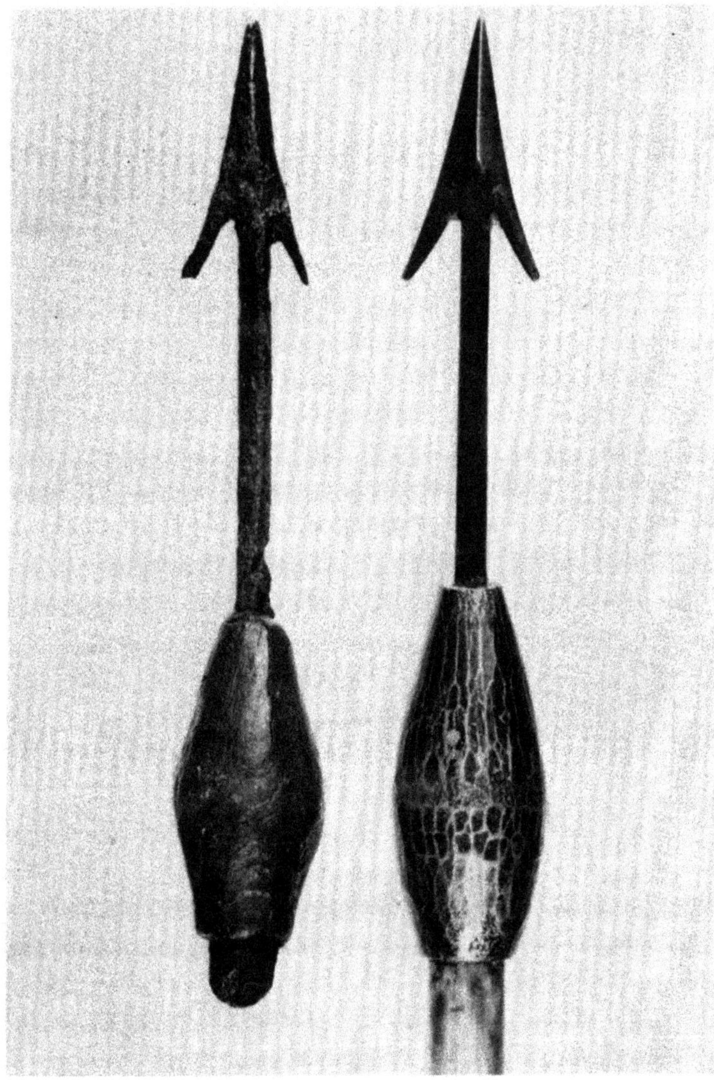

PLATE XI. Plumbata from Wroxeter, left original, right replica. Scale 1:1

The neutron radiographs showed that the plumbatae were of split socket type, with some wood remaining in situ.

The plumbatae of the Anonymus were equipped with feathered flights like an arrow, and his description and pictures provide the only evidence for this feature. In order to try to establish the length of the wooden shaft behind the weight on the Wroxeter plumbatae and whether they were also fletched, a number of replicas were made at the Tower of London Armouries through the good offices of the late Mr. Russell Robinson. Throwing experiments carried out at the Tower and in Worcester showed that it was difficult to achieve distances of more than about thirty yards with a shafted and fletched plumbata simply thrown by hand. However, with the aid of a string the range was increased to seventy or eighty yeards. No doubt with practice it would be possible to increase this distance.[8]

Department of Extramural Studies,
University of Birmingham,
P.O. Box 363,
BIRMINGHAM, B15 2TT.

## NOTES AND REFERENCES

1. Veg. iii 14.
2. Thompson, 68.
3. J. Curle, A Roman Frontier Post and its People (Glasgow, 1911), Pl. XXXVIII.
4. E.g. Excavations at Dura-Europos, Report of the Sixth Season (Yale, 1936), Pl. XXIV.
5. The latest buildings from among whose foundations the plumbatae were found are not closely datable. They contain nothing but Roman material but a good deal of this may be residual. The number of periods of occupation at present recognised between the destruction of the basilica and the latest buildings suggests a date in the late fourth or early fifth centuries for the beginning of the last period of occupation. A tombstone of an Irish chieftain, Cunorix Macus Maqui Coline ('Cunorix, son of the son of the Holly'), dated by Professor Kenneth Jackson to 460-480, Ant. J. 48 (1968), 296-300, is the latest find from the city and implies an Irish mercenary force operating there in the later fifth century. However, plumbatae were weapons of the regular Roman army and should date from the early fifth century at latest.
6. Publication forthcoming.
7. J. Corbet Anderson, The Roman City Uriconium at Wroxeter, Salop. (London, 1867), Pl. XIII.
8. For a fuller account of the plumbatae replicas and experiments made with them, see Ant. J. 54 part ii (1974), 175-277.

# PLUMBATAE–A NOTE ON THE METHOD OF MANUFACTURE

by David Sherlock

The recent finding at Burgh Castle Saxon shore-fort (*Gariannonum*) of an iron spear-head, almost certainly from a *plumbata*, not only adds to the known distribution of this class of weapon but also tells us more about its method of manufacture.[1] The Burgh Castle iron is 15.8 cm long, about 25% larger than the best-preserved of the Wroxeter examples which measures 11.8 cm from the point to the further end of the leaded section, and larger still than those from Richborough (*Rutupiae*)[2] and Doncaster (*Danum*)[3] forts, though smaller than the 19th-century Wroxeter find[4] and a comparable one from Castell Weissenberg on the German *limes*.[5]

Neuron radiographs showed that the Wroxeter *plumbatae* described in the report above were of "split socket type, with some wood remaining *in situ*". To this method of joining shaft and head may be added a detail that can be obtained from the illustration of the 19th-century find which is described by the author as follows: "(The weapon) is flat-headed, and about 10 inches long; the nail which formerly held its long wooden shaft still remains."[6] (Fig. 14b) On the other hand, the Burgh Castle example, lacking its lead weight and wooden shaft, shows an alternative method of combining the three parts of the weapon, as shown in Fig. 14a. Which of the two methods was the more effective is hard to judge. The Burgh Castle method would be speedier to make but without a nail the shaft might work loose more easily, especially if the surface of the wood was charred by the moulten lead. In the Wroxeter method there is less wood exposed to contact with the lead and a nail would have held the wood firmly in the iron socket.

Inspectorate of Ancient Monuments
Fortress House, 22 Savile Row
London W1X 2HE

## NOTES AND REFERENCES

1. *Proc. Suff. Inst. Arch.* 34, Part 2 (1978), 141-3.
2. *Richborough* 4 (Soc. Antiq. Research Rept, 1949), 152 and Pl. LIX, nos. 295-296.
3. *Current Archaeol.* 3 (1971-2), p. 274, illus.
4. Anderson, *op. cit.* (p. 99, n. 7).
5. *Castell Weissenberg* (O.R.L. 72, Heidelberg, 1914), S. 42, Taf. VIII, Nr 88.
6. Anderson, *op. cit.*, p. 74.

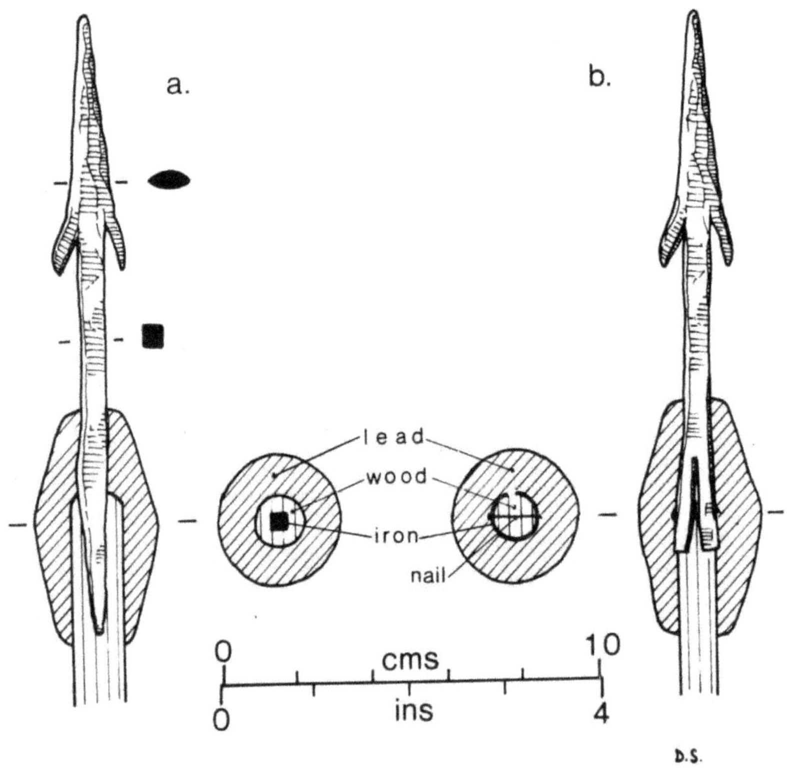


FIG. 14 Two suggested methods of manufacturing *plumbatae*:
*a*. The Burgh Castle iron inserted into end of wood
*b*. The same iron adapted with split socket end encasing wood and held with nail, in the manner of the Wroxeter examples

# FOURTH-CENTURY UNDERWEAR
# WITH SPECIAL REFERENCE TO THE THORACOMACHUS

by J. P. Wild

My subject is the thoracomachus, and I shall consider both the Anonymus' verbal description of it, (XV) and the subtilis pictura which he provided to illustrate his meaning (Part 2, Pls. XVI, XVII).

Briefly put, the thoracomachus is a shirt of sheeps' wool felt, designed to protect the chest of the wearer against the weight and friction of his armour, and against cold. In wet weather the Anonymus recommends that a rainproof sheepskin tunic should be worn over it. The author thus cites or illustrates a number of other items of clothing and armour, some of which also deserve comments.

Firstly, the name thoracomachus. It was invented in all probability by the Anonymus himself. Greek scientists frequently coined new technical terms when the need arose - for instance, Theophrastus' ἀείφυλλος, 'evergreen'. Latin was a less accommodating language, and the Anonymus' choice of Greek was in keeping with Roman practice. θωραξ , 'breastplate' or 'breast' and μάχομαι,'fight', are the constituent elements of the term.[1] The author seems to construe the compound as 'chest-protector' (ex tuitione corporis: XV. 4), although this is not the only sense which the Greek could bear.

The thoracomachus is to be made from soft wool felt (de coactili...de mollibus lanis) to fit the human body (ad mensuram...pectoris humani XV. 2.). Felt was a familiar material in antiquity and had a wider range of uses than knitted or woven textile fabrics.

The most famous archaeological finds are probably the Scythian felt coverings, hangings and clothing from the barrows at Pazyryk in the Altai region, dating to the fifth and fourth centuries B.C. The fabrics are decorated with brilliantly coloured applied leatherwork and embroidery.[2] More prosaically, in the classical world one may cite the embroidered saddle-cloths listed in the Edict of Diocletian or the felt covering on the plywood shield lost by a Celtic mercenary in the Fayum in the third or second century B.C.[3]

Felt linings were usual for many items like helmets and slippers,[4] but few items of actual clothing were made from it. The most obvious exceptions are felt caps, which we shall consider later, and udones, long felt socks.[5]

Sheeps' wool, because of the scaly outer structure of the wool fibre and its crimp, is ideally suited to making felt, but we hear of goats' hair felt (cilicia) and hares' hair felt in Roman times, too.[6]

To make felt from sheeps' wool moist heat, friction and pressure must be applied - hence the name *coactile*.[7] The nomads of the Central Asian Steppes still make felt by the traditional method. They moisten the scoured wool with whey and spread it out in a thick layer on a blanket which they then roll and unroll repeatedly. Eventually the wool fibres become a stable, compact, fabric.

Roman methods were industrialised.[8] A wall-painting in Pompeii outside the workshop of Verecundus, a *coactiliarius*, shows his employees, the *lanifricarii*, kneeding rolls of hot, moist wool fibres in shallow sinks supported on trestles.[9] Surplus water drained back into a boiler between them. Several other houses in Pompeii containing free-standing boilders have been claimed as the premises of felt manufacturers.[10]

Felt was not often used for body-garments in classical antiquity. Even soft wool felt would not have been particularly comfortable next to the skin. Moreover, lice were one of the facts of life - as Aristophanes assures us - and there was no point in offering them a secure home.

The *thoracomachus* as it is revealed in the extant manuscript illustrations (Part 2 Pls. XVI, XVII) was a long-sleeved shirt of standard late-Roman pattern. The style was called a *strictoria* (*sc*. *vestis*) on account of its tight sleeves and comparatively restricting cut. In the fourth century it was on sale in linen, sheeps' wool, rabbits' or hares' wool, silk and a silk-and-wool union fabric.[11] The *orantes* on the wall-paintings of the Christian chapel at Lullingstone give us a good impression of the style and appearance of the *strictoria*.[12]

The illustrator of the *Codex Spirensis* drew on the right-hand side of his vignette two shirts hung on the Roman equivalent of the modern coat-hanger. One is the *thoracomachus*, the other the sheepskin overtunic. Both have fairly wide sleeves, but the foot-soldier on the left-hand side who is wearing the *thoracomachus* has the regulation tight sleeves.

The copyist who worked on the Bodleian manuscript C reproduces more convincing details than the rest (Pl. XVI.*a*.). He gives the *thoracomachus* a rounded neck-aperture, edged with a dark-coloured band. This might be interpreted as a cloth binding to prevent the raw edge of the felt from chafing the soldier's neck; but it is more likely to be meant as embroidery or tapestry weaving in tune with contemporary fashion. The cuffs are finished off in a similar manner, but there are no additional tapestry-woven *emblemata*;[13] for most of the garment would have been invisible, especially if a chain-mail *clibanus* were to be worn over it. If the *thoracomachus* had gone into production in the state *gynaecea*,[14] it would have been equivalent to the *strictoria indictionalis*, the shirt cut to the taxation-office norm, and decoration would surely have been minimal.[15]

To what extent was the thoracomachus a new idea? The *Anonymus* claims that he is advocating an old device, credit for which he gives to 'the ancients'.

In describing the Athenian capture of Sphacteria in 425 B.C. Thucydides states that the Spartans' πῖλοι did not keep out the Athenian arrows.[16] The word is unfortunately ambiguous; it means any item of felt, a cap, leggings or some other garment.[17] In the context editors translate 'felt cuirass', and certainly that makes better sense than cap or leggings. If we accept this interpretation - some authorities do not - then a passage in Pollux[18] where mention is made of the fame of Spartan πῖλοι may refer to the same felt cuirass. The Notae Tironianae[19] list the shorthand sigla for thorax and coactilis close together, but this may be coincidental.[20] No extant source apart from the de Rebus Bellicis mentions a felt shirt to be worn under armour.

We are on firmer ground in stating that felt was a recognised strategic material long before the time of the Anonymus. Odysseus' boars' tusk helmet was lined with felt (μέσσῃ δ' ἐνὶ πῖλος ἀρήρειν [21]) and the metal helmets worn throughout the classical period must regularly have had a felt lining, if only as a shock absorber. One of the Newstead parade helmets, for example, dating to the late first century A.D., was found to contain remains of a felt lining.[22] The Greeks went to considerable lengths with padding to minimise the effect of blows on a helmet-encased skull.[23] E. A. Thompson drew our attention to Ammianus' companion who had the forethought to wear a pad of old rags (cento) under his helmet, perhaps as a supplement to an existing felt lining.[24]

Charioteers wore a felt cap, the pilleus, as a crash-helmet. A letter written in the third or fourth century by a gladiator to his trainer at Oxyrhynchus bears a request for a pilleus, a pair of μαμυλιῶνες (probably a waistcoat) and a pair of felt bands.[25] Another papyrus, from Antinoe, dated to the fifth century, illustrates a group of charioteers in their racing colours. They have yellow pillei on their heads and their bodies are protected by what could well be the μαμυλίωνες and the felt bands mentioned in the letter.[26]

When Caesar's men were troubled by the arrows of the Pompeians during the siege of Dyrrachium, we are told that they made themselves tunics and coverings of an unspecified nature out of felt, old rags and hide.[27] This was a reaction to an emergency, not normal practice.

In the field Roman soldiers wore their armour, whatever type it might be, over a woollen or a linen tunic - sometimes over both. An additional leather jerkin was worn under chain-mail and scale-armour, as Trajan's Column and archaeological discoveries indicate. Felt does not seem to have played a part in this scheme; but the uses of felt were familiar to everyone.

The Anonymus neatly anticipated the practical objection that a rain-sodden thoracomachus would be extremely uncomfortable. Today, the tribesmen of the Steppes grease the felt segments which make up their tents to render them waterproof, but this would have seemed a desperate remedy to a Roman soldier. Our author suggests that a tunic de Libycis ... pellibus (XV.3), worn over the thoracomachus, but presumably under the breastplate, would solve the problem - if the soldier could have fought under such a weight of protective covering! Vegetius tells us that soldiers in the later fourth century had become oversensitive about wearing unnecessary gear.[28]

The manuscript illustrations show the overtunic as the same shape as the thoracomachus. The soldier standing left does not appear to be wearing an overtunic. On the other hand the copyists responsible for the illustrations in MS.C in Oxford and MS.M in Munich (which contains two sets of illustrations) disagree on whether the Codex Spirensis showed a soldier in full kit or not. So there is some confusion.

The peculiar virtue of Libyan skins (I assume goatskin or lambskin) is obscure.[29] But leather was certainly the main material used in waterproof clothing in antiquity.[30] We can be sure that the hair or wool would have been removed from the Anonymus' Libycae pelles; for the emperors Honorius and Theodosius II in December of A.D. 416 issued a particularly bad-tempered constitution against long hair (maiores crines) and fur coats (indumenta pellium) in the city of Rome.[31] So much for the thoracomachus itself.

The Anonymus quotes a number of other items of clothing and equipment which made up the soldier's accoutrements. His picture shows them, too.

At the foot of the tree in the centre of the scene is a pair of socci, slippers. We know from other sources that these were of leather, sometimes gilded and dyed.[32] They were house-shoes, not military boots, and so seem strangely out of context here.

On the tree behind them hang the iron greaves, ocreae ferreae, prescribed by the Anonymus. At the top of the tree is what can only be the galea, helmet, but the copyists do not seem to have understood. it. The artist who drew the figure in the text of MS.M (Part 2 Pl. XVII.a) thought that he was portraying the helmet from the front, and clearly depicts what he took to be cheek-pieces. He may have been right. The Bodleian MS.C has less detail, but shows what may be enlarged hinges for the cheek-pieces on either side of the helmet. (Pl. XVI a, cf.b). At any rate the Codex Spirensis appears to have reproduced a picture of the late Roman Gardehelm or Spangenhelm.[33]

Finally, an item which the Anonymus had drawn, but did not mention in his text. The standing soldier wears a cap on his head, the pilleus. We have already had cause to mention the charioteer's pilleus; but this form of headgear is found in many other contexts in classical antiquity.[34] As its name implies, it was of felt, probably mould-made. However, the Edict of Diocletian lists in the chapter on leatherware a 'sheepskin with the wool left on for making a pilleus', which cost twice as much.[35] A fleecy pilleus, I think, is what the Anonymus' picture was meant to show.

Soldiers are characterised in late antique art by the pilleus, which has the form of a fez. The soldiers seen arresting St. Peter on late Roman sarcophagi have a pilleus more often than not.[36] Even the so-called tetrarchic figure on the Piazza Armerina mosaics has a cap of this type.[37] A variation on the same theme is provided by the apparel of the guard watching over the tomb of Christ on a fifth-century ivory diptych in the Trivulzio Collection.[38] Vegetius throws a little light on this military fashion. He comments that until recently all soldiers had worn leather pillei, called Pannonian caps, so that when they came to put on a helmet, they would already be accustomed to having a weight on their heads.[39]

It is greatly to his credit that the our author realised that a comfortable soldier is more likely to be an efficient soldier. Humble though it seems, the thoracomachus deserves at least as much attention as his more spectacular inventions.

Department of Archaeology,
University of Manchester,
MANCHESTER M.139.PL.

## NOTES AND REFERENCES

1. Liddell-Scott-Jones s.v. θώραξ, μάχομαι. Thoracomachus is ignored by those who have written on Greek terms in Latin.
2. S. I. Rudenko, Frozen Tombs of Siberia (London 1970), 83ff.
3. Ed.Diocl. (Lauffer's edition) VII, 52, 53; Germania 24, (1940), 106ff.
4. J. P. Wild, Textile Manufacture in the Northern Roman Provinces (Cambridge, 1970), 120ff.
5. Udones: Martial, xiv.140; Pollux, Onom. x.50; Ed.Diocl. VII.47; P. Ryl. IV.627.35; Dig. XXIV.2.25.4 (Ulpian).
6. Goat: Martial, XIV.140; hare: Wild, op.cit. (n.4), 120 B97.
7. For production methods: CIBA-Review 11, no.129 (November 1958).
8. Wild, op.cit. (n.4), 60.
9. V. Spinazzola, Pompei alla luce degli scavi nuovi di via dell'Abbondanza (anni 1910-1923) 1, (Rome, 1953), 194, Tav.d'agg.2 (I do not think that the vestiarius Verecundus (CIL IV.3130) need be the same man, pace Spinazzola); W. O. Moeller, 'The lanifricarius and the officinae lanifricariae at Pompeii', Technology and Culture 7 (1966), 493ff.
10. W. O. Moeller, 'The Felt Shops at Pompeii', American Journal of Archaeology 75 (1971), 188ff.
11. Ed.Diocl. XIX.10,11,18,73a; XX.1a,2; XXII.2,9,14; XXVI.13ff.
12. HMSO, Lullingstone Roman Villa, (1962), 31; B.Jb 168(1968), 185 Bild 12,2, and P193.
13. For example on the mosaics of the villa near Piazza Armerina: W. Dorigo, Late Roman Painting (London 1971), colour Pls. 99 and 101.
14. Latomus 26 (1967), 648ff., R. Goodburn and P. Bartholomew (eds.). Aspects of the Notitia Dignitatum, B.A.R. Supplementary Series 15 (Oxford 1976), 51-58.
15. Ed.Diocl. XIX.2; XXII.2; XXVI.28; JRS 60 (1970), 128.
16. Thucydides iv, 34.
17. Liddell-Scott-Jones s.v. πῖλος ; PW(-K) 20 pt.2, 1330.
18. Pollux, Onom. i, 149.
19. Notae Tironianae xcvii. 26.28.
20. Müller evidently thought that it was significant: Berliner Philologische Wochenschrift 31 (1911), 232 (with incorrect reference taken from TLL).

21. Iliad x.265.

22. Wild, op.cit. (n.4), 102 A53.

23. I am grateful to my colleague, Dr. Alastair Jackson, for information on this question.

24. Amm. xix. 8.8. A cento is technically an amalgam of old rags, not felt.

25. P. Oxy. 2598; the Latin mamillare cited by the editors ad loc. is female attire: Martial, XIV.66.

26. Journal of Egyptian Archaeology 17 (1931), 1ff., Pl. 1.

27. Bell. Civ. iii.44. Body armour of quilted and rolled cotton was worn in Mexico and adopted by the Spanish conquistadors as protection against darts and arrows. Transactions of the American Philosophical Society, New Series 47 (1957), 361. I owe this reference to Mr. J. Forde-Johnston.

28. Veg. i.20.

29. Most pelles registered in the Edict of Diocletian (VIII.6ff.) are without provenance; for sheep in North Africa: Classical Journal 49 (1953-54), 157ff. Dio Chrysostom (xxxv.12) mentions a goat-skin cloak σισύρα ); for a discussion of types of leather: A. Gansser-Burckhardt, Das Leder und seine Verarbeitung im römischen Legionslager Vindonissa, (Basel, 1942), 7ff.

30. Martial, xiv.130. Woollen capes with a long nap were waterproof, too: Antiquity 37 (1963), 198.

31. Th. XIV.10,4.

32. Ed. Diocl. IX.20a, 21ff.; Isid. Etym. xix.34.12.

33. H. Klumbach, Spätrömische Gardehelme, Münchner Beiträge zur Vor- und Frühgeschichte 15, (Munich, 1973); for literature on Spangenhelme: Proceedings of the Society of Antiquaries of Scotland 94 (1960-1), 184ff. (I owe this reference to Mr. Richard Higginbottom).

34. PW(-K) XX, 2.1328 s.v. pilleus.

35. Ed. Diocl. VIII.15.16 (Greek and Latin entries). I believe, pace Lauffer, tha bellearis here (for vellearis) means 'with the fleece (vellus) left on', not 'mit gerupfter Schafwolle'.

36. F. W. Deichmann (ed.), Repertorium der christlich-antiken Sarkophage I, Rom und Ostia (Wiesbaden, 1967), nos. 6, 7, 11, 17, 20 et passim.

37. W. Dorigo, op. cit. (n. 13), Pl. C1.

38. R. Delbrueck, Die Consulardiptychen und verwandte Denkmäler (Berlin, 1929), no. 68.

39. Veg. i.20.

# SIEGE-CRAFT FROM THE SIXTH TO THE TENTH CENTURY

by David Hill

It seems certain that the text and accompanying illustrations of the *de Rebus Bellicis* had already been copied at least once before being copied yet again by the compilers of the *Codex Spirensis* in the late ninth or early tenth century. The patrons who ordered the text to be transcribed and the pictures to be carefully reproduced, may have been inspired by mere antiquarianism, but their interest in the weird and wonderful machines of the Anonymus could have had a more practical aspect. Were the Carolingian Franks familair with the use of the *ballista*, for example, or was the whole concept of torsion artillery alien to them? In trying to answer such questions it is worth looking at accounts of sieges in the Early Middle Ages, for here if anywhere we might expect to find sophisticated military equipment being employed both in attack and defence.

The accepted view of the period of Western European History from the 'withdrawal' of the legions to the arrival of the Normans is so tinted with 'primitive', 'tribal' and 'barbarian' overtones that we are sometimes likely to miss altogether important aspects of the period. Thus it is possible in England to see the whole Anglo-Saxon period as a mere prologue to the "real" Middle Ages, this is done simply by labelling the period as *Pre-Conquest*. This tendency should be resisted by recognising the length of the period, and its variety, and by recognising the sophistication of some of the European states at this period. In the Frankish lands the Kings came to see themselves as the inheritors of the *romanitas* and finally of the title and trappings of the Empire. The period is complex and many sided. The influence of Rome disappeared gradually, the techniques of the High Middle Ages have their roots before 1066. Fortifications of various kinds are to be found in the period from the collapse of the Western Empire to the rise of the Normans, it might be a mistake therefore to accept the received opinion that siege machines and siege warfare are anachronistic, that they disappear completely to be re-invented in all their ramifications in the sieges of the Normans. It is not to be accepted without discussion that for six hundred years there were no siege machines because the people were waging a hack-and-bash type of warfare.

The centre of the western Christian kingdoms was the Kingdom of the Franks, the old province of Gaul. The siege-engines and the whole science of poliorcetics may well have been absent from many of the minor kingdoms for most of the centuries but in Gaul the Franks waged war which continually turned on the cities and fortresses. Where warfare is constantly hinging on the possession of fortifications there must be sieges, and it follows logically that there may well be siege engines and military engineering. There was never a long enough period for the science to disappear completely from the time of Belisarius and his defence of Rome onwards. The sources are not good but if we follow the campaigns of the Franks we can demonstrate in every period some sieges and some form of military engineering and siege engines.

In the epic siege of Rome (March 537-March 538) we see all of the Roman siege engines and expertise in use, and some evidence of it in use on the side of the Goths. Belisarius used a great deal of intelligence and most of the classic engines. When Convenae in Southern Gaul was besieged in 585 we have an account from the History of the Franks, of Gregory of Tours, which includes descriptions of fascines for filling ditches, stones which are hurled and vessels of burning pitch and of grease which are prepared. Siege engines are in evidence:-

> 'The fifteenth day of the siege dawned and Leudegisel was preparing new engines for the destruction of the city. There were wagons fitted with rams covered with wattle-work and planks, under which troops could go forward to destroy the walls. But as soon as they came near they were overwhelmed with such showers of stones that all fell who approached the wall'.[1]

The mention of fortresses and particularly of the sieges of cities continues throughout the following centuries but rarely are they explicit. Perhaps the most fruitful source, which follows on from Gregory of Tours is the Fourth Book of the Chronicle of Fredegar with its Continuations.[2] An example of the type of warfare that might be used comes from one of the many internecine wars, and in the entry for 639 we learn:-

> 'Observing this, Radulf put up a wooden stockade round his position on a rise above the river Unstrut...and when he had assembled every possible man from all parts, fortified himself with his wife and children within this field-work to withstand a siege....'

The siege itself is not dwelt on, but here is an example of a field fortification in use. It may well be argued that the most sophisticated forms of military artifice were re-learnt at regular intervals from the Byzantines, and there is some support for a view that siege machines of the most advanced types appear in Francia and north-western Europe after prolonged contact, particularly in warfare, between Frank and Eastern Roman, and more particularly between Frank and Lombard, a race particularly influenced by contact with imperial domains in Italy. Of course it may simply be that we are reading too much into the fact that references become fuller as we move towards the eighth and ninth centuries. Fredegar's account of a southern siege by the Franks shows much of the military art, in 737:-

> 'Once more, the mighty race of Ishmael, who are now known by the outlandish name of Saracens, rebelled and burst across the river Rhone... Childebrand lost no time in bivouacking in the approaches and surrounding countryside of this excellently provisioned city (Avignon), and in setting about his preparations for the forthcoming engagement until the arrival of Charles, the Great Commander, who forthwith shut up the city, erected siege-works and tightened the blockade. Then as once before Jericho, the armies gave a great shout, the trumpets brayed and the men rushed into the assault with battering rams and rope ladders to get over walls and buildings....

later Charles crossed the Rhone... He invested its famous capital, Narbonne itself. He threw up lines on the banks of the Aude in which he installed offensive armament of the battering type. He then continued his lines in a wide sweep around the Saracen emir Abd ar-Rahman and his viziers, investing them so. He added carefully constructed works at intervals.'

The method of circumvallation etc. is clear, the types of engine is maddeningly inexplicit. In the siege of Bourges (762) we are similarly, and vaguely told:-

> 'He (Pippin) then surrounded the town with a strong fortification, with siege-engines and all manner of weapons so that no one would have dared to leave it or enter it, and he also built a rampart.'

We have now reached the threshold of the reign of Charlemagne; sources increase and become a little more explicit. The capitularies lay down the equipment for the army on campaign and include references to what would appear to be engines but which are explained, according to the orthodox view, as some sort of staff-sling.[3] This translation would, of course, fit in with a view of the period which dubs it 'primitive', ballistae are not primitive or tribal so we have the staff-sling. But how would this square with the annal for 776 in the Annales regni Francorum[4] which tells not of the Franks but the Saxons besieging them? they had not learnt well from their opponents, but they had learnt:-

> (The Siege of Syburg) 'When they (the Saxons) failed to talk the guards into surrender, as they had those in the other castle, they began to set up war machines to storm the castle. Since God willed it, the catapults which they had prepared did more damage to them than those inside. When the Saxons saw that their constructions were useless to them, they prepared faggots to capture the fortress in one charge. But God's power, as is only just, overcame theirs.'

There are many references in the ninth century to sieges and their machines. They appear in Lombardy and around the periphery of the Empire. The impression continues that these machines are built on the spot by a skilled artificer rather than transported, although the essentials were perhaps always carried. Rather than list all the references we should consider the most famous of all the sieges, that of Paris.

The best description of this siege, or any ninth century siege is in a curiously neglected document, Abbo's Bella Parisiacae Urbis. This is not the place for an extended discussion of the authenticity of the text. It should be sufficient to say that many modern texts and translations are to be found in France and a translation has recently appeared in Danish, indicating a continental acceptance of Abbo's work, at least as literature.[5] The lack of an English translation and discussion is remarkable. The long poem had reached England by the middle of the tenth century, circulated widely and a partial translation into Old English appeared. Since then English historians have ignored the source. The actual events were well known in England and appear both in the Anglo-Saxon Chronicle and in Asser. From the internal evidence we know that Abbo was an eye-witness and that the poem is partially a eulogy of Eudes, Count and later King, and can be confidently ascribed to the year 897. It recounts in considerable detail and appalling poetry the events of the Siege of Paris from 885 to 886. If we accept the advice of another poet who was used to writing in praise of another medieval King:-

> 'We find the best evidence in the poems which were offered to the kings themselves or to their sons; we take everything for true which is found in their poems about their journeys or battles. It is the way of skalds, of course, to give the most praise to him for whom they are composed, but no one would dare tell the king himself such deeds of his as all listeners and the king himself knew to be lies and loose talk; for that would be mockery but not praise.'[6]

The whole poem is poor and obscure latin but it does contain the most detailed descriptions of the military arts of the period. As their is no translation in English it may be worthwhile to quote a few of the relevant passages _in extenso_. It has often seemed surprising to many that it is the Danes as well as the Franks who are using these advanced tactics, but the account of the Siege of Syburg in 776 must alert us to the swift interchange of ideas in time of war, for then it is the Saxons who are attempting to use poliorcetics. A more striking example is to be found in Ceasar's Gallic Wars where the Gauls are shown to be capable after a very few years contact with Roman attacks to carry on a very respectable siege.[7]

The initial attempts of the Danes to capture Paris in the autumn of 886 brought Eudes to the fore (i. 99-102):-

> 'If anyone tried to cut into the foot of the wall with iron picks, he waited upon them with oil, wax and pitch. This mixture, liquefied, in a burning furnace burnt the Danes' hair and made them tear it from their heads.'

As the _coup de main_ fails the Danes institute a regular seige, although, as Paris is on an island they never completely circumvallate it. Siege Engines appear (i. 205-220):-

> 'Meanwhile the Danes undertook the making of some veritable monsters, and to put them on sixteen wheels, a most extraordinary sight, the like of which the Franks had never seen before. Three of these grouped together, made from strong oaks, were each placed over a battering ram, hidden by a raised roof. There was enough space in the secret depths of their flanks to hide sixty helmeted men. Yet they only succeeded in building one such machine, of a fairly ample size. The second was finished, and they were working on a third, when a missile was fired arfully and with force from a _ballista_, it arrived and killed the two machine-builders together. So they were the first to suffer the death they were preparing for the Parisians. Soon the evil couple perished, mortally wounded by the single blow. They had gone to the extent of setting up a thousand awnings on elevated supports which they covered with skins taken from the back and neck of young bulls. They acted like shields and each was capable of sheltering three of four men; this is what a scholar would call a _mantelet_ or screen.'

The sophistication that is claimed for some of the mechanical devices is surprising (i.235-6, 295-300):-

> 'Thousands of grenades of molten lead were thrown into the town. On the bridges powerful catapults mingled with the watchtowers.' '.... The Danes carried the canopies already mentioned to the tower, These were made from wood from the forest and from the skins of young slaughtered bulls. Thanks to which some of them passed the night making war whilst others slept, and they made circular openings in the canopies through which to fire feathered arrows impregnated with gall.'

Finally the great attack was launched (i.356-374):-

> 'They put their battering rams into action; they set one of them to the east facing the tower; another to the north, against the gates and a third was put to the west. The Franks prepared some heavy pieces of wood each with an iron tooth in the end, so as to damage the Danish machines more quickly. With coupled beams of the same length they built what are commonly called Mangonels, machines for throwing vast stones, which would blast the lowly race of barbarians often blowing out their brains, crushing crowds of them and their shields. Not one shield that was hit did not break; not one unfortunate who was hit did not die. In vain their columns which nothing could resist tried to fill in the ditches, they could not fill in a single one. They tried to destroy the tower with blows from the battering ram, but were unable to move them forward on flat ground...'

The Siege lasted for months, with intermissions, and we read of wagons filled with a slow burning load of turf used against the gates of one tower, of fireships towed against the bridges and so on. All this against a missile fire which is continuous and, to use modern parlance, both ingoing and outgoing (ii.239-244):-

> 'And shields groaned beneath the weight of heavy balls of lead and massive stones. This was the way the Danes fought all the time. The Franks replied in a lively way, aiming stones and ballista bolts and swift arrows at the savages. Missiles were flying everywhere and crossing each other in the air.'

There are other references to poliorcetics and machines in this period at Angers, Sens and other towns, unfortunately apart from the fact that both Frank and Dane are employing them there is little detail to be gained from the several Chronicles of the period.

It is difficult to reconstruct from these few references any clear idea of what was happening. It would seem that the towns sometimes had fixed defensive machines, both ballistae and mangonels, and that siege trains were not attached to the armies but a few skilled artificers travelled in the host. Certain parts of the machines were probably present in the trains the rest being made as time went on and a formal siege was decided on. Battering rams, ballistae, fascines, mantelets, fire-wagons, ladders and circumvallation were general, on occasion the water supply was diverted and strict blockades enforced. Such techniques may have been practised ever since Roman times or knowledge of them may have come from repeated contact with the 'Greeks' of the Eastern Roman Empire. For example in the next century during the difficult siege of the Saracens in Fraxinetum on Gaul's Mediterranean coast, Hugh, Count of Arles and King of Italy, sent to Constantinople for fireships. With the Byzantine Navy attacking from the sea using the famous Greek Fire, and Hugh's men from the land, the Saracens came to terms. There is also a possibility that some of the mechanics involved may have been revived from late Latin sources which returned to general circulation in the Carolingian Renaissance. Vitruvius contains information which is of sufficient detail to be of practical use and we know that copies of his work were made and in circulation throughout this period in the West, less certainly Vegetius would appear to have had some currency and the only copy of the Anonymus appears to have made in a Carolingian provenance. How far the science spread outside the borders of the Frankish Empire is even less certain, both Saxons and Danes used the machines when necessity forced them, and it should be remembered that the Danes had their own contacts with the Eastern Mediterranean, but to the north the evidence for such machines in the Kingdoms of the West Saxons is poor. The Anglo-Saxon Chronicle is unspecific when talking of the details of warfare and no extended account exists for any one siege. But it should be recalled that in 885 the Danes attacking Paris had come directly from besieging Rochester.

Certain it is that members of the West Saxon monarchy, clergy and nobility were in close and frequent contact both with northern France and with Rome. And it is precisely this class that were amused by, and must have had the ability to solve, the Anglo-Saxon Riddles. Riddle Seventeen has been translated variously but there is virtual unanimity in modern editors as to the interpretation of the subject: the definitive edition by Krapp and Dobbie sees the answer as probably '*ballista*'[8] while Tupper, in the fullest discussion of the riddles,[9] points out that the riddle is part of a pair with 'bow' and almost certainly is '*ballista*'. In translation it reads:-

'I am the protector of my flock,
fast strengthened with wires and filled inside
with noble treasures. In the day-time I often
spit forth deadly spears; I have the greater success
the fuller I am. My master beholds
How missiles of war fly from my belly.
Sometimes I begin to swallow black and dusky
weapons of war, bitter arrows,
spears dreadful and venomous. I have a good stomach,
and within it an excellent treasure, dear to gallant men.
Men remember what passes through my mouth.'[10]

Whatever the doubt about riddle seventeen however, there can be no doubt about the weapon intended in riddle 53 and the translation reads:-

'I saw a tree, gleaming with branches,
stand erect in a forest. That tree was joyful,
a growing timber. Water and earth
fed it well, until, when old in years,
it became changed, deeply gashed

by torment, dumb in bonds,
fettered over its wounds, adorned in front
with dark trappings. Now it clears a way,
by the might of its head, for another,
dangerous warrior; they often together
fiercely pillaged the hoard. Active and quick
was the second, if the first,
his comrade in dangers, must venture into danger.'[11]

The answer is taken as 'battering-ram' in all recent authorities and would give us evidence for the knowledge of a pair of siege machines at a date of *circa* 970. Together wich such defensive features as the multiple ditch systems, at Lydford, Wareham and Christchurch, caltrops from a Saxon context at Ipswich[12] and the use of siege works and circumvallation by the Danes at London and perhaps Rochester[13] a more sophisticated military picture must be accepted.

## NOTES AND REFERENCES

1. O. M. Dalton, *The History of the Franks* (Oxford, 1927), 314.
2. J. M. Wallace Hadrill, *The Fourth Book of the Chronicle of Fredegar* (London, 1960).
3. *MGH Leges* 2.1, No. 77, section 10. (Aix, 802/3). The word used is *fundibulae*. The stones to be used as missiles were to be transported by twenty pack-horses.
4. B. W. Scholz, *Carolingian Chronicles* (Ann Arbor, 1970).
5. For a text available in larger libraries see *MGH scriptores rerum Germanicarum in usum scholarum separatim editi* 1 8$^{o}$, (Hanover, 1871).
6. *Heimskringle* or *The Lives of the Norse Kings*, preface. See the edition by E. Mansen, (Cambridge, 1932), p xxxvi.
7. A. L. F. Rivet, 'Hillforts in Action' in D. Hill and M. Jesson (eds.) *The Iron Age and its Hillforts* (Southampton, 1971), 189-202.
8. G. H. Krapp and E. V. K. Dobbie, *The Exeter Book* (New York and London, 1936), 332.
9. F. Tupper, *The Riddles of the Exeter Book*, (Boston, 1910)..
10. W. S. Mackie 'Notes on the Text of the Exeter Book', *Modern Language Review*, (1934), 107.
11. *ibid.*, 145.
12. *Proc. Suff. Inst. Arch.* 29 (1964) 274 no. 3 and Fig. 54.
13. Cf. The Anglo Saxon Chronicle entries for the year 1016 (London) and 885 (Rochester).

# THE ANONYMUS IN THE LATER MIDDLE AGES

by Henry Hodges

This paper is concerned with the question of what effects, if any, the writings of the Anonymus had upon military technologists of the later Middle Ages. This may seem a strange question to put, but since at least two copies were made of the work in the first half of the fifteenth century one is entitled to ask who were its readers, and why it was found worth while to reproduce the original manuscript.

In his recent monograph entitled The Renaissance Engineers Bertrand Gille[1] has conveniently summarised the history of notebooks on military technology of the late Middle Ages, beginning with Guido da Vigevano (c. 1328) and ending with the notes made by Leonardo da Vinci in the period 1485-1488. Vigevano's treaties, written for Philip V de Valois, was designed as a military manual for crusaders. In the same vein was the Bellifortis of Konrad Kyeser (c. 1405) dedicated to the Emperor Ruprecht of the Palatinate. The seventy-odd years between these two works was a critical period in the development of military technology: Vigevano knew nothing of firearms, Kyeser's machines literally bristle with guns. This trend was carried still further in the decades of the Hussite wars, and the work of the Anonymus Hussite (c. 1430), which may be seen as Europe's first handbook of a resistance movement, illustrates the first true cannons. It is evident that both these manuscripts were available to the Italian engineers Taccola and Valturio since many of the illustrations in the latter authors' works are direct copies from those of the German school. Little is known of Taccola save that his manuscript, De machinis libri X, must have been produced c. 1450. Roberto Valturio, better known as an architect, finished his treatise on military matters for Sigismundo Malatesta in c. 1455. A third Italian engineer, Francesco di Giorgio Martini, completed his manuscript on military technology between the years 1470-1480, while Valturio's work first appeared as a printed edition in 1472.

This extremely brief survey, which omits many of the minor writers on the subject, is sufficient to show the keen interest that existed from the mid-fourteenth century onwards in manuals of military technology. Although dedicated to members of royal or noble houses, there seems little doubt that they were written for the benefit of other engineers and military leaders. One must suppose that many engineers had their own collections of private notes just as a chef de cuisine today will keep his own recipe book. It is important to grasp that, quite apart from a few original ideas, each collection contained a high proportion of devices borrowed from a common pool. It is equally important to realise that many of the inclusions were made, not because of their utility or practicability, but for the sake of completeness.

We may now ask which of the machines described by the Anonymus were most likely to have found their way into these notebooks. There are basically four such devices: the catapults, the inflatable pontoon bridge, the paddle-driven boat and the chariots. The remainder of the equipment is of such a fundamental nature that it would be impossible to trace in this context.

The first on this list, the catapults, can be dismissed quite rapidly. By the middle of the thirteenth century torsion artillery had been largely replaced as a means of throwing heavy missiles by the mangonel, quickly superseded by the trebuchet.[2] These machines were essentially heavy pivoted levers to one end of which was attached a sling in which was placed the missile. In the case of the mangonel the other end of the lever was equipped with hauling lines operated by a team of men; while in the trebuchet[3] these ropes were replaced by a heavy counterweight, normally of stones or soil contained in a large sack, and which was therefore capable of being varied according to circumstances. As Lynn White has pointed out,[4] the trebuchet was far superior in matters of range and accuracy than any torsion-operated artillery of the early Middle Ages, especially once it had been established that rounded shot of standard weights could be flung repeatedly at the same target area with a reasonable chance of hitting it. Thus, had the writers of the later Middle Ages seen the Anonymus' accounts of catapults they might well have meant nothing to them.[5] At best they would be recognized as outmoded pieces of equipment, more difficult to construct than the trebuchet and less effective.

The only similar device to survive throughout the Middle Ages was the cross-bow,[6] largely as a hand-held weapon. It is certainly significant that when Leonardo da Vinci drew a machine to throw a heavy spear he saw it not as a ballista but as a scaled-up version of the hand-held cross-bow of the period.[7] To it he attached a carriage of large wheels, but as far as one can detect provided no means of training or elevating the weapon. His sketch was presumably either a preliminary study of a subject on which he decided to do no further work, or a record culled from some other source simply out of interest.

The inflatable pontoon bridge described by the Anonymus seems at first sight to be a possible source of inspiration for a similar kind of bridge depicted by Vigevano in which one could argue that the inflated skins of our Author have been replaced by casks.[8] If so, this would be a surprising modification since the whole contraption was supposed, once broken down to its components, to be readily portable on pack animals and one would imagine that skins, folded and ready for inflation, would have been far more easily carried than barrels. On the other hand it might have been that casks, initially containing victuals, would have been available on any crusading expedition, while skins might not have been to hand. Even so, elsewhere in his text, Vigevano describes the use of inflated skins as floats for swimmers crossing rivers.[9]

Kyeser, too, has his pontoon bridge, which appears to have been a series of rafts joined by lattices that could be closed or extended depending upon the width of river to be spanned.[10] This contraption seems to owe little to Vigevano's bridge and nothing to that of the Anonymus. Furthermore, inflated skins for swimmers are described frequently by the other writers on military technology. Even in this century rafts made up in the manner described by the

Anonymus and skin floats have been reported in use widely throughout Asia from Mesopotamia to China.[11] The use of such rafts and floats in the Middle Ages was even more widespread, in which case their inclusion in texts on military technology need have nothing to do with the work of our Author. Thus Leonardo writes, (When the Spaniards, Scythians and Arabs wish to build a bridge very quickly, they bind hurdles made of willow to bags or wine-skins of oxhide, and so cross safely).[12]

Boats propelled by paddle-wheels occur in virtually all the texts on military technology from Vigevano to Leonardo, but none show the ox-powered capstan drive of our Author. Instead, the paddle-boats are small vessels intended to be manually operated, and that illustrated by Vigevano incidentally shows one of the first uses in Europe of the double crank handle.[13] Needham has recently drawn attention to the early development of similar paddle-boats in China,[14] but as in the case of cross-bows cited above it is highly doubtful that any question of diffusion is involved.[15] Indeed a major difficulty that stands in the way of any discussion of the evolution of the paddle-boat lies in the fact that a perfectly reasonable prototype had been evolved in the late Roman period and continued in use throughout the Middle Ages. This was the floating mill, basically a moored boat equipped with an undershot water-wheel that operated the mill-stone set amidships.[16] In this case the turning wheel provided the power, but by reversing the process one had a paddle-boat. It seems highly probable that this was one source of inspiration for our Author's paddle boat, the other being the ox-driven water-raising wheel, common to all areas of irrigation farming, that provided the power.[17] The fact that Vigevano and later authors on military technology[18] used only muscle-power strongly suggests that the paddle-boats of the late Middle Ages were a separate and distinct development that owed nothing to our Author's work.

So far we have drawn a blank. There is nothing positive to suggest that our author's text was known to military writers of the later Middle Ages, and there remain only the chariots to consider, (Part ?, Pls. X-XV). In a period that was to see the final development of heavy cavalry and plate armour it is hardly surprising that the late Medieval writers have nothing to say about chariots of the kind illustrated by our author. Admittedly, Vigevano has a lumbering ox-drawn wagon fitted with fixed scythe blades[19] and this was copied by later writers in modified forms; but these machines have no bearing upon our author's chariots, which could be more correctly described as 'defensively equipped cavalry'.

Evidence that the text and accompanying illustratio of the Anonymus were known to Leonardo da Vinci comes in the form of a most surprising series of sketches (Pls. XII-XIII)[20] made in the period 1485-1488. Here we see our author's scythed chariots re-vamped to suit current ideas of Renaissance engineering. The single draught-pole of the currodoepanus clipeatus (Part 2, Pls. XIV, XV) has been replaced by shafts or traces while the 'automatic flogging device' has given way to a fearsome group of whirling balls and chains (Pl. XII). The latter in all probability owe their design to the ball-and-chain governors to be seen as compondents of hand-mills in the works of the Anonymus Hussite, Taccola and Giorgio:[21] but the inspiration may well lie in the illustrations of our author's text, for the dangling bells (?) attached to the horseman's 'wings' certainly look at first sight like a pair of balls and chains.

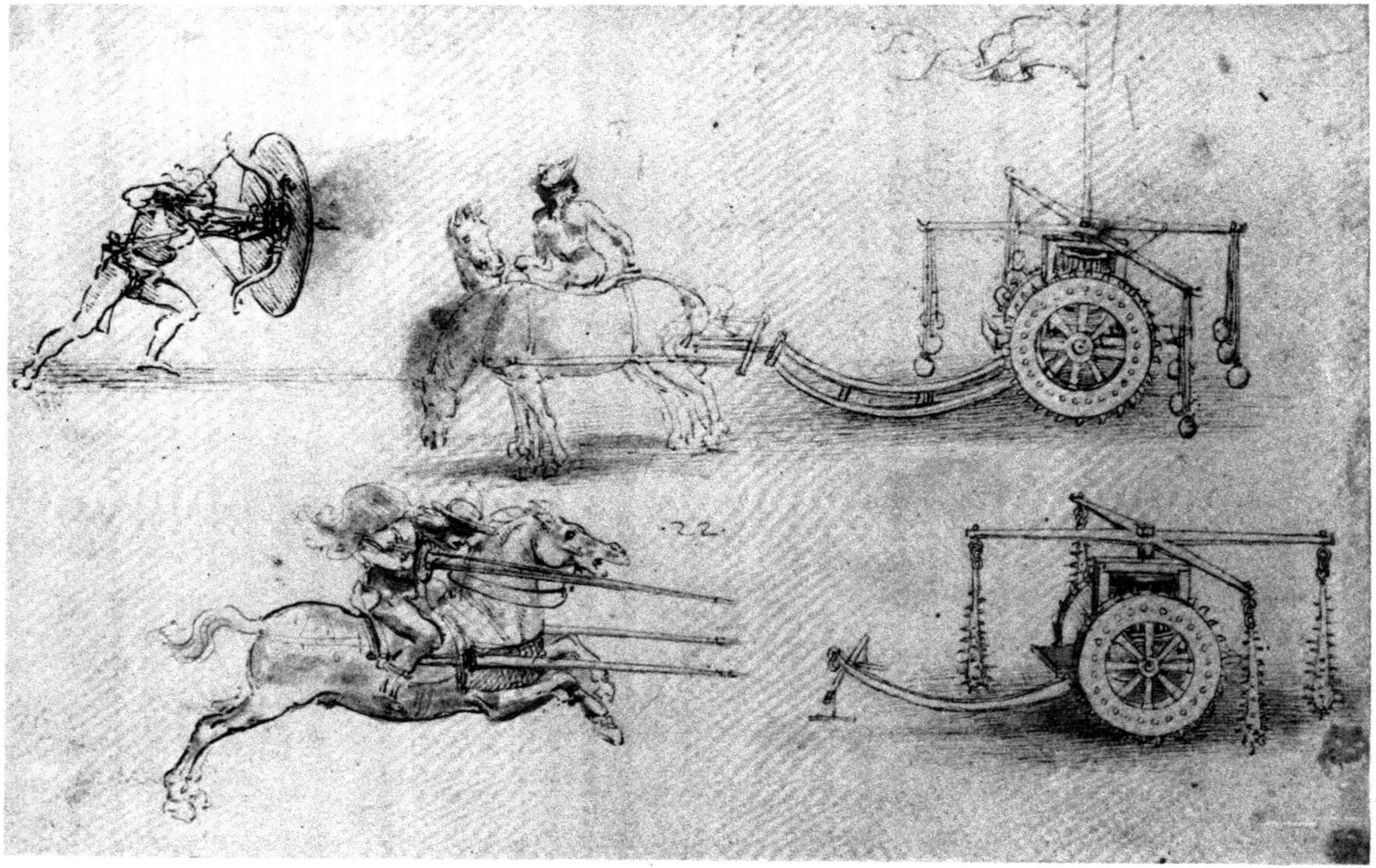

PLATE XII. Leonardo sketches, Windsor Castle Royal Library. Inventory No. 12,653. Reproduced by gracious permission of Her Majesty the Queen

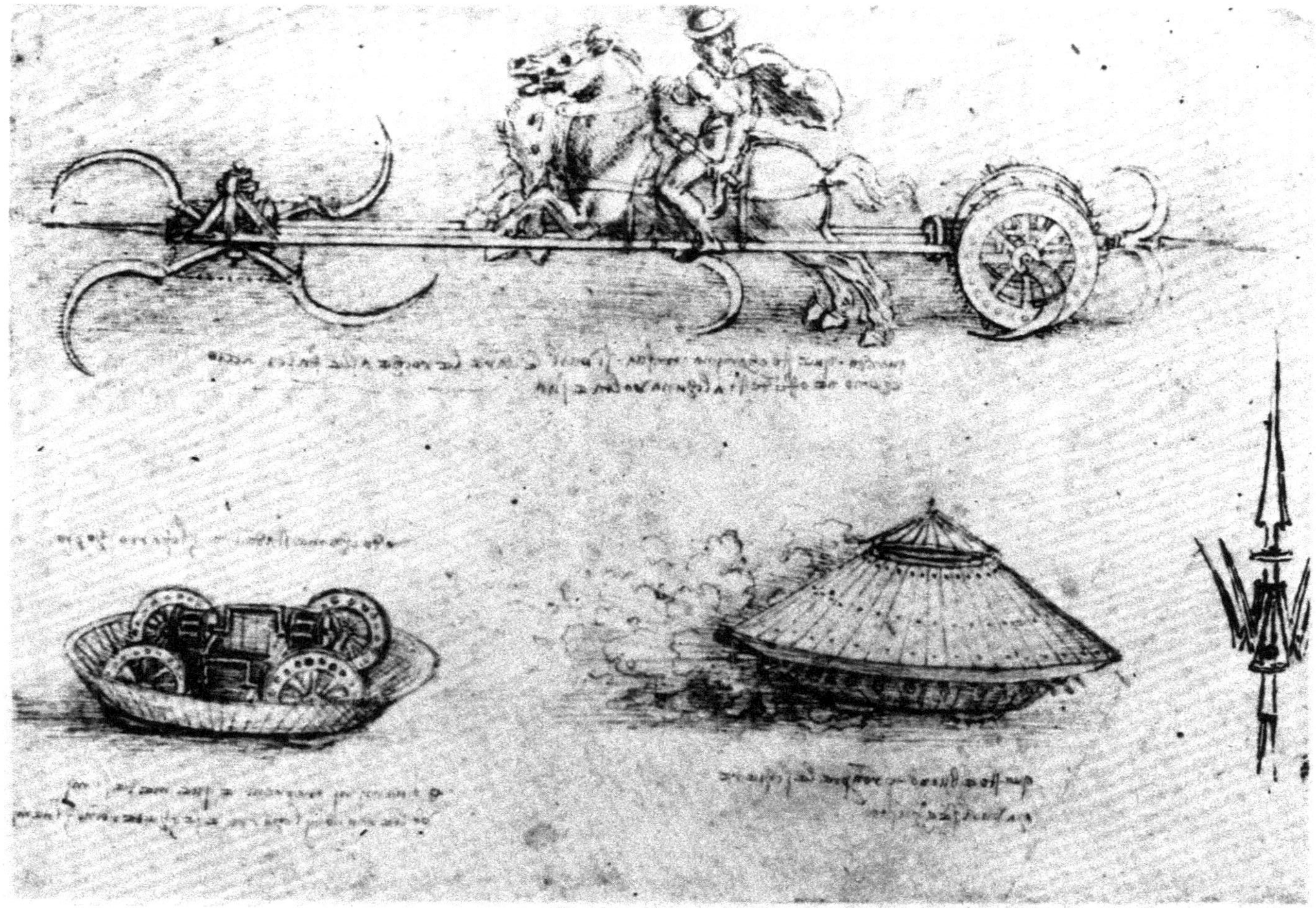

PLATE XIII. Leonardo sketches, British Museum, Cat. No. 107

Furthermore, the cloaks worn by some of Leonardo's riders, truly unnecessary encumbrances, may well be what Leonardo made of the 'wings'. The (Currodrepanus (Part 2, Pls. X, XI), in which the two riders are seen to hold their lances at exactly the same angle, may have inspired the curious and totally impracticable arrangement of fixed lances on Leonardo's lone horseman (Pl. XII bottom left); while the sketch of an archer on the same page (Pl. XII top left), in which he is seen to be shooting through the central opening of a shield, may have been incluenced by our author's illustration of the clipeocentrus (Part ?, Pls. VII, VIII).

Leonardo's text makes his indebtedness to the Anonymus even clearer. "These chariots armed with scythes were of various kinds," he wrote, "and often did no less harm to their friends than to their enemies, for the commanders of the armies thinking that by their use they would throw confusion into the ranks of the enemy caused panic and destruction among their own forces....As a protection against them the Romans used to scatter iron caltrops, which brought the horses to a halt and made them fall to the ground in pain, leaving the chariots unable to move."[2?] Elsewhere he writes, "When passing through its own ranks it is necessary to raise the machinery that works the scythes to prevent them harming anyone."[23] The mention of "various kinds" of chariot armed with scythes; of the need to raise the scythes when passing through one's own forces; and of the use of caltrops (compare the plumbata tribolata, DRB X) leaves no reasonable doubt as to which Roman author served as Leonardo's authority. It is quite typical of Leonardo, too, to have dealt with our author's chariots in the manner he did. For him machines had to be functional, and the fact that these chariots had no bearing upon the military situation of his day was irrelevant. Dürer's picture of a landsknecht tells us the truth about the cavalry of the period: Leonardo's sketches belong to a fairy-tale world of Ancient Rome put into modern dress.

It is not diffuclt to understand why the de Rebus Bellicis of the Anonymus was copied in the first half of the fifteenth century. Clearly it belonged on the shelf with Vigevano, Kyeser and the others, if only for the sake of completeness. It was required reading because it was history and because it dealt with topics that suited the mood of the Renaissance. As a practical guide to matters of military technology it had, as on the day it was written, no function at all.

Department of Art History,
Queen's University,
Kingston,
Ontario,
CANADA.

## NC 'ES AND REFERENCES


1. Gille, Renaissance Engineers (London, 1966).

2. A MS in the National Library, Vienna, dated 1415 has an illustration of a petrary or onager which is torsion operated (Gille, op.cit. (n.1), 26; 57). Machines of this nature were, nevertheless, a rarity at this time.

3. Gill, op.cit. (n. 1), 26; C. Singer et al. (eds.), A History of Technology, Vol. 2 (Oxford, 1956), 644.

4. L. White, The Expansion of Technology (London, 1969), 32.

5. This is assuming that the Anonymus artillery was torsion operated. In fact his descriptions are so unclear that even this is not certain.

6. Writers on the subject of medieval technology disagree as to the continued use of the cross-bow, either in the mounted or handheld form, in Western Europe after the Roman period. By some authorities the lack of written or pictorial evidence for its use prior to the first crusade is taken as evidence of its total absence, and its 're-introduction' from China has been laid at many doors. Recently the Kara Khitai have been credited with this supposed diffusion (J. Needham, Science and Civilization in China, Vol 4, 3 (Cambridge, 1971). However, the very different string-retaining and trigger mechanisms of the Chinese and European cross-bows argue against such diffusionist theories. Compare Singer, op. cit. (n.3), 723 with H. Hodges, Technology in the Ancient World (London, 1970), 226.

7. Milan, Cod. Atlant. 53v.

8. Gille, op.cit.(n.1) 29.

9. ibid., 29.

10. ibid., 62..

11. Hodges, op.cit. (n.6), 152.

12. Paris MS. B62v.

13. Gille, op.cit. (n.1), 28; Singer, op.cit. (n. 3), 651.

14. Needham, op.cit. (n.6). Vol. 4, 2 pp. 413-35.

15. See footnote 6 above.

16. Singer, op.cit. (n. 3), 607-8.

17. ibid., 675; Hodges op.cit. (n. 6), 193; T. Schiøler, Roman and Islamic Water-Lifting Wheels (Odense, Denmark, 1973).

18. Gille, op.cit. (n. 1), 62.

19. *ibid.*, 31.

20. *Cf*. A. E. Popham, *The Drawings of Leonardo da Vinci* (London, 1946), Pls. CCCVIII-XI.

21. Gille, *op.cit.* (n. 1), 72, 111.

22. E. MacCurdy, *The Notebooks of Leonardo da Vinci* (London, 1938), 185.

23. *ibid.*, 212.

The late C. E. Stevens' extempore summing up at the Anonymus Seminar was transcribed from a tape recording and is here reproduced with the minimum of editorial alteration as a tribute to a great scholar and a remarkable man.

## SUMMING UP

by C. E. Stevens

I started thinking about the Anonymus when it was almost impossible to get a copy of his work at all and when it went on the record as what Seeck called "die Denkschrift einer verrückten Projectenmachers". Well now I think that we have something extremely interesting and suggestive for students of the later Roman empire trying to think Anonymus out. It comes from all sorts of different angles. It comes from Stephen Johnson on Frontier Policy, it comes from John Peter Wild on the Thoracomachus and it comes from Mark Hassall on ballistae—that, in fact, the Anonymus, when we look at him, was inventing very little indeed. He was telling the emperor what he ought to have known about already. I now completely accept that the Annonymus' "mile-castles" need not be, and probably were not the mile castles on Hadrian's wall as I once had suggested,[1] but were simply what Valentinian was building at the time. I cannot believe that Valentinian read the Anonymus and said, "Good gracious, I've got to do this" because, as Stephen Johnson said, it was all in the line of Military thinking.

And that brings me on to consider, what was the Anonymus really up to? Because every single person had their terms of reference, I would like rather to try to tie their contributions together with something with which I think they would all agree with. Let me look at the background of Valentinian-Valens with special reference to what the Anonymus wrote.

In the preface he said what I would expect him to say with the language of a frightened man: "I may yet get into trouble over this". Now that was only too common at that time. The emperors, from Constantius onward, were all, I think, under desperate strain. If they were not ninnies, they were very, very tempersome men. Theodosius, Valentinian, they all came like that. It was a very, very dangerous book to write and he said so.

Now what I think he was up to, was really what was so extremely well brought out for us by Patsy Vanags and Richard Reece; he really wanted to relieve the financial difficulties with special reference to soldiers and favourites. Ammianus knew the latter. Aurelius Victor finished his Caesars by pointing out what a lot of waste goes on favourites under a bad emperor, but then Constantius was conveniently dead. Seldom do we have criticism of a living emperor; Nobody dares to criticize the military policies of an emperor unless he can in fact get him because he was on, so to say, the losing side. Yet it is just the questions, "How much more the army costs us. How can we economise in the army?" that concerned the Anonymus.

It was safe for Lactantius, as a Christian under the Christian Constantine, to point out: this military expenditure is all very well but what a lot it costs. Lactantius could say that. I do not worry that I have to go back to

Lactantius to look at my parallel for the Anonymus. Now, who was the Anonymus? He was not a man, as he very carefully makes clear, who was in the Civil Service. He hadn't climbed the ladder which put him in charge of *quaestiones*; he was not involed in the giving of justice; he was not a provincial governor; and he wasn't "educated in eloquence", he wasn't an *advocatus fisci*. He did not climb the civilian ladder at all. Let us first keep that in mind.

What are the implications of this? As a civilian but a private citizen, was he not really a rather clever man, getting the emperor's ear for something that was safe, to disguise putting across something that was very dangerous? Thompson has said that it was naive to say that we should get our taxation straight. Agreed, it was naive but, it was *desperately* dangerous. It was saying that an emperor could't pick the right ministers. I remind myself that perhaps the more or less contemporary *Historia Augusta* put up a man who said to Hadrian: "Domitian may have been a bad man but he picked well" - with the inference that a good man could pick badly. It was a dangerous thing to say. I can think of the people who made the complaints against Valentinian and Valens - who got into trouble when they did. These were the emperors who had picked Count Romanus. A desperately dangerous thing to say: "You're picking badly". How did he get round it? He got round it, I think, by putting these military machines in and for the most part putting things in about which the emperor would say: "Well, that's rather clever of him, I thought of that myself".

And when he does try technical improvements, they are silly improvements. (The paddle steamer, and I would say the treadmill attached to the *ballista fulminalis*, incidentally, seem to indicate a familiarity with mines - I would like you to remember the wheels at Rio Tinto and even one at Dolaucothi in Wales). They are included but it is interesting to see how he starts with the financial-political stuff, and then how, after the machines, he has rather an interesting lead back by the mile forts, to finance again: you can save money by employing quasi-civilian *burgarii* in fifties or hundreds, who get their orders from the civilian Praetorian Praefect in the Theodosian Code. He led up to the military machines and then led away from them finishing with something which, in a sense, ought to have been at the beginning, that is to say the codification of Roman Law and all the difficulties that the civilians had got into.

I will make you a guess at the end. His military improvements are very slight and rather silly. He was not in fact a military man at all. He ought not, as I have said, have been in the civil service if he was not a hightly educated man. He was, perhaps, a *curialis* and I am sure that an eastern origin is right, otherwise, if he had been a westerner, quite, simply, he would have mentioned not the Danube but the Rhine. He lived somewhere near the Danube, I would guess. And I would guess too, that he had seen these military machines because he was somewhere where there were *comitatenses* attached. Whether or not they actually had offensive machines such as the Anonymus suggests, the dangers the Romans faced required defensive ones which the barbarians didn't have, but the Romans certainly possessed. We think it hightly likely that the bastions of Roman towns were built so as to mount ballistas upon them. Defensive machines are important. In including the offensive machines he is really rather silly. What are they going to attack a town for? The barbarians did not live in towns. If he had seen anything like them in reality they may very well have been kept on the establishment of comitatensian units of the field army, offensive machines to be used against Roman towns if, perhaps, there were a civil war.

That is the picture I meant: that he is a man very interested in, very thoughtful about the dangers - "The budget doesn't balance, the taxes are too high, too much goes on the soldiers, on the army. Can we think it out? The emperor ought to do something about this." Like Iphicles who made complaints against Probus of fiscal oppression in Illyricum in 375. Perhaps he was a *curialis* under Probus. "Very dangerous thing to do. Let us put the machines in. I have seen them myself. (*Comitatenses* were billeted in towns.) I'll think out a clever thing or two. Perhaps this will go far." Well, I think the fate of the Anonymus was the fate of the man who pushed the tank up to the War Office. The War Office man wrote on it: "The man's mad". But it got onto the file. It got onto the file.

He couldn't do anything about his worry, though I think myself that he *could* have got it both ways. He felt, correctly, that it would have benefited the *collatores* if you got the military onto the land and made them pay taxes there. I think this was a modest idea that he could have got from his own region, that if you could put them on emphyteutic leases, they might pay a low rent but they'd pay something. They could be called *collatores*, if they did pay. It was a neat half-way house if they paid something at least.

The man's mad. But I don't think we'd have had his pamphlet somehow if he hadn't been clever enough at least to keep his head on his shoulders. The nicest thing about this conference is to show that he is worth studying. He is not a "verrückter Projectenmacher" and neither are the lecturers verrückte Vortragende, nor is this very old hand in the business, who once published something on the Anonymus, and I now know it's not worth the paper it was printed on!

## NOTE

1. C. E. Stevens, 'A Roman Author in North-West Britain'. *CW*[2]1. (1951), pp. 70-79.

# PART TWO

# ANONYMI AVCTORIS LIBELLVS

# DE REBVS BELLICIS

## A TREATISE BY AN UNKNOWN AUTHOR

## ON MILITARY MATTERS

edited, translated and presented
with commentaries on the text, language and style

by ROBERT IRELAND

- ἄκρον δὲ τοῦτον εἶναι τελεστικόν. καὶ δύνασθαι δέ φησι καὶ ὑπισχνεῖσθαι αὐτὸν χωρὶς ὁπλιτῶν κατὰ βαρβάρων ἐνεργεῖν.

Olympiodoros, frg.38

PRELIMINARY

nunc tandem. *Hastily conceived and painfully gestated, this edition of* de rebus bellicis *is finally,* obstetricante me ipso, *delivered as the sequel to a set of learned papers whose appearance has been delayed for five years by my reluctance to take leave of an old and valued companion. To the authors of these papers I owe my apologies; to the reader, my explanations; and to my friends, my thanks. I have kept my work on the text very largely to myself,* neque si male cesserat olim decurrens alio, neque si bene; *and the component parts of my contribution were pulled together at so late a stage as to render the advice of external authority - even if it had been sought - unserviceable*: si autem uel pro mei abiectione uel pro uitiosi sermonis rusticitate placet nulli, memet tamen ipsum iuuat quod feci. *One portion of this edition , however, has been seen by other eyes than mine, and I must thank Paul Collins for reviewing my translation, and 2/Lt P.A. Fox R.A. for correcting its use of military terminology. The translation and parts of the interpretative work were drafted during a period of residence at the Fondation Hardt in Geneva, and I am very grateful to Professors Eric Turner and Olivier Reverdin, Mme Suzanne Moor, M. Bernard Grange, and their staff, for making my stay at La Chandoleine possible, delightful and productive. And I must record some purely personal obligations: to Mark Hassall, for imposing the present task upon me and for subsequently applying no more than the necessary pressure; to Mark Lowe, for limitless supplies of coffee; to Mark, Corinne, Adam, Daniel and Magdalen Gretton, for the weekend in Suffolk during which* (*I hope without their knowledge*) *the book was finished; to Royston, Jo, Timothy and Matthew Ball, for refuge at Wargrave during crucial stages of the typing; to my parents, who did not ask awkward questions, and to my brother and sister, who did. In technical matters I am indebted to Ian Marriott, for invaluable help with* the computer procedure employed in producing the concordance to the text, and to Moira Mackenzie, who pasted up text and translation with unerring skill and a good-humour unshaken even by a last-minute disaster involving a picture-caption.

The typography of the edition is my own: I have to thank Professors D.M. Wilson, G.P. Goold, E.W. Handley and J.E.C.T. White for successive grants towards the purchase of the IBM machine and the stock of types with which these pages were executed. The reader has therefore to contend not only with an aggressive and despotic editor but with an ambitious and untrained typist as well. He will probably find occasion to blame me in both my capacities, and I shall defend myself in both with the words of Foulke Robartes: Who faulteth not, liueth not; who mendeth faults is commended: The Printer hath faulted a little: it may be the author ouersighted more. Thy paine (Reader) is the least; then erre not thou most by misconstruing or sharpe censuring; least thou be more vncharitable, then either of them hath been heedlesse.

As for the editing: there are certain obvious and unassailable corrections which every post-1952 text of our treatise must incorporate, and the user of Thompson's edition will find here much that is already familiar. But he must also be prepared to find some novelties, even where he thought that the solution of difficulties had been found: I have not hesitated to reconsider the text at all its points of weakness, whether previously suspected or not. In the commentaries I have very often used the language of hypothesis, sometimes of intimidation: the editor of a difficult text in a treacherous tradition, first edited by Siegmond Ghelen and in modern times all but neglected, has little choice in the matter. The art of boring may consist in saying everything: but it is equally true that the only really interesting questions are those to which no answer has yet been found. The obvious will be found here juxtaposed with the speculative; explorations into the inner nature of a fourth-century style co-exist with a catalogue of the most elementary uses of the ablative. It was necessary to do all this now, because none of it had been done before: and I have tried to leave the text and its literary interpretation at a point, not where all future work is precluded, but from which it may usefully start.

CONTENTS

*An analysis of the contents of sections* 1-5 *is printed on pp.* 193-196

# 1 TEXT

## A NOTE ON THE PRESENTATION OF THE TEXT

The following typographic conventions have been adopted here:

*italics* in the text, angle brackets < > in the critical notes, signify supplements, *italic asterisks **** a postulated lacuna;

words within square brackets [ ] are editorially deleted;

the roman asterisk * precedes a correction or supplement (or a lacuna assumed) by the present editor;

obeli † † surround irremediable corruptions.

To permit exact reference to the text the traditional chapter-division has been maintained but the sentences within chapters have been re-numbered throughout.

The illustrations in the text are reproduced at full size from the *editio princeps* of 1552. Froben's engraver naturally transferred to his blocks a direct copy of the drawn illustration before him, and the printed impressions are, in consequence, laterally reversed with respect to the originals. No attempt has been made to alter this by photographic means in the present edition. A necessary change, however, has been ventured on in reproducing the illustrations to cc. x-xi (the *plumbatae*) and xvi (*ascogefyrus*), both of which Froben's compositor locked into the text *upside-down* - a pardonable error in both cases, but an error nonetheless, and one which is here corrected. And to comply with the lateral reversal of the printed image of the blocks for cc. viii-ix and (as restored) x-xi, the items within the captions to these blocks have been transposed so that the correct name appears above each device; the text chapters of course follow in their transmitted sequence. (The *editio princeps* avoids the problem here by ranging the names vertically over the centre of the cuts). Finally, in the interests of typographical consistency, the original letterpress dropped into the illustration for c. i has been replaced by labels set in the type used for this edition.

# DE REBVS BELLICIS

## praefatio

1 caelesti semper instinctu felicis rei publicae uestrae commoditas,
sacratissimi principes, opportunis est suggerenda temporibus, ut
2 diuina consilia diuinis successibus conualescant. unde pro ingenii
facultate unum capitulum de largitionum utilitate in hoc libello
composui, non quod istud tam inmensae utilitati sufficiat sed ut
ex hoc mediocritatis meae documento praemisso in reliquis utili-
3 tatis possit fides ostendi. proinde, ne promissioni fides accom-
modata uelut fallacem in posterum grauet, pro conscientia uerita-
tis ipse mihi poenam praemii loco, si promissio secus cesserit,
posco, *nec me laus aut *praemium* prosequatur cum abunde sit in hac
4 parte indignationem pro audacia declinare. sed fas erit rei publi-
cae praesulem a priuato desiderata cognoscere, cum rerum utilitas
5 interdum eum lateat inquirentem. ergo non numquam inuitandi sunt
qui recte quicquam sentire fuerint approbati: nam ut ait optimus
orator, ad illum maior pars hominum decurrit quem ingenio natura
6 donauerit. in qua re est considerare semper quid unus quisque magis
sentiat quam loquatur; constat enim apud omnes quod nec summa nobi-
litas nec opum affluentia aut subnixae tribunalibus potestates aut
eloquentia litteris acquisita consecuta est utilitates artium, in
quibus etiam armorum continetur inuentio, sed ingenii tantum modo
magnitudo, quae uirtutum omnium mater est, naturae felicitate sub-
7 nixa. quod quidem sine personarum electione uidemus accidere: nam
cum barbarae gentes neque facundia polleant aut dignitatibus illus-
trentur, minime tamen a rerum inuentione, natura opitulante, haben-
8 tur alieni. quam ob rem, clementissimi principes, qui gloriam bonae
opinionis perpetua felicitate diligitis, qui romano nomini debitos
affectus propagatis in filios, respicere dignemini quae nostris sen-
9 sibus commoda prouidentia diuinitatis intulerit. uniuersis igitur
seu militiam clementiae uestrae tractantibus seu otio priuato con-
tentis uel terrae cultoribus siue negotiatoribus mercium lucra
tractantibus pro saeculi uestri felicitate gaudentibus, conseque-
mini ex hoc opere commoda singulorum, quorum species diuersis
titulis, opportunis quibusque locis, oratio subiecta testabitur.
10 referemus enim quem ad modum, remissa tributorum medietate, in
robur proprium prouinciarum cultor habeatur: unde etiam, **in*
ratione exactionum cessante contumelia, limitum solitudinem erec-
tis castrorum munitionibus incola securus illustret: quo etiam
pacto auri argentique modus sine dantium poena duplicetur: uel
quo argumento extra solitam largitatem cumulatus honoribus miles
11 exultet. his etiam adnectenda credidimus quae bellorum necessi-
tatibus terra uel mari in acquirendis uictoriis procurentur; ex
quibus, fastidii leuandi gratia, pauca machinarum inuenta refe-
12 remus. docebimus igitur uelocissimum liburnae genus decem nauibus
ingenii magisterio praeualere ita ut hae per eam sine auxilio
13 cuiusquam turbae obruantur. in terrenis uero congressibus talis
est excogitata sollertia ut equus, siue aciem rupturus inuadat
siue fugientibus ingerat se, tali arte muniatur ut semet ipsum
uerberans sine cuiusquam magisterio efficiat magnas hostium
14 strages. in fluuiorum quoque difficultatibus exquisita est noui

pontis inuentio minime subuectioni onerosa; hunc enim pontem
amnibus paludibusque necessarium perpauci homines aut quinqua-
15 ginta fere numero iumenta portabunt. ex his igitur quae rettu-
limus, iam futuris ut ego arbitror designatis, pace uestra
dixerim, magnum uobis munus concessu diuinitatis apporto, asse-
rens prouidentia pietatis uestrae armorum uigorem et cunctam
rem publicam praedictis remediis subleuandam: quae quidem non
ignota sunt proximis uestrae clementiae, quos alia plura sol-
16 licitant a nobis aliena. uerum quia illos multa occupatos ef-
fugiunt, otio persuasus, non adeo a rerum commoditatibus pere-
grinus, utilia uestrae felicitati undique redacta conferre
17 gestiui. si quid uero liberius oratio mea pro rerum necessitate
protulerit, aestimo uenia protegendum, cum mihi promissionis
implendae gratia subueniendum est propter philosophiae libertatem.

COMMODAE AVCTORITATIS VARIAE PRISCORVM MONETAE

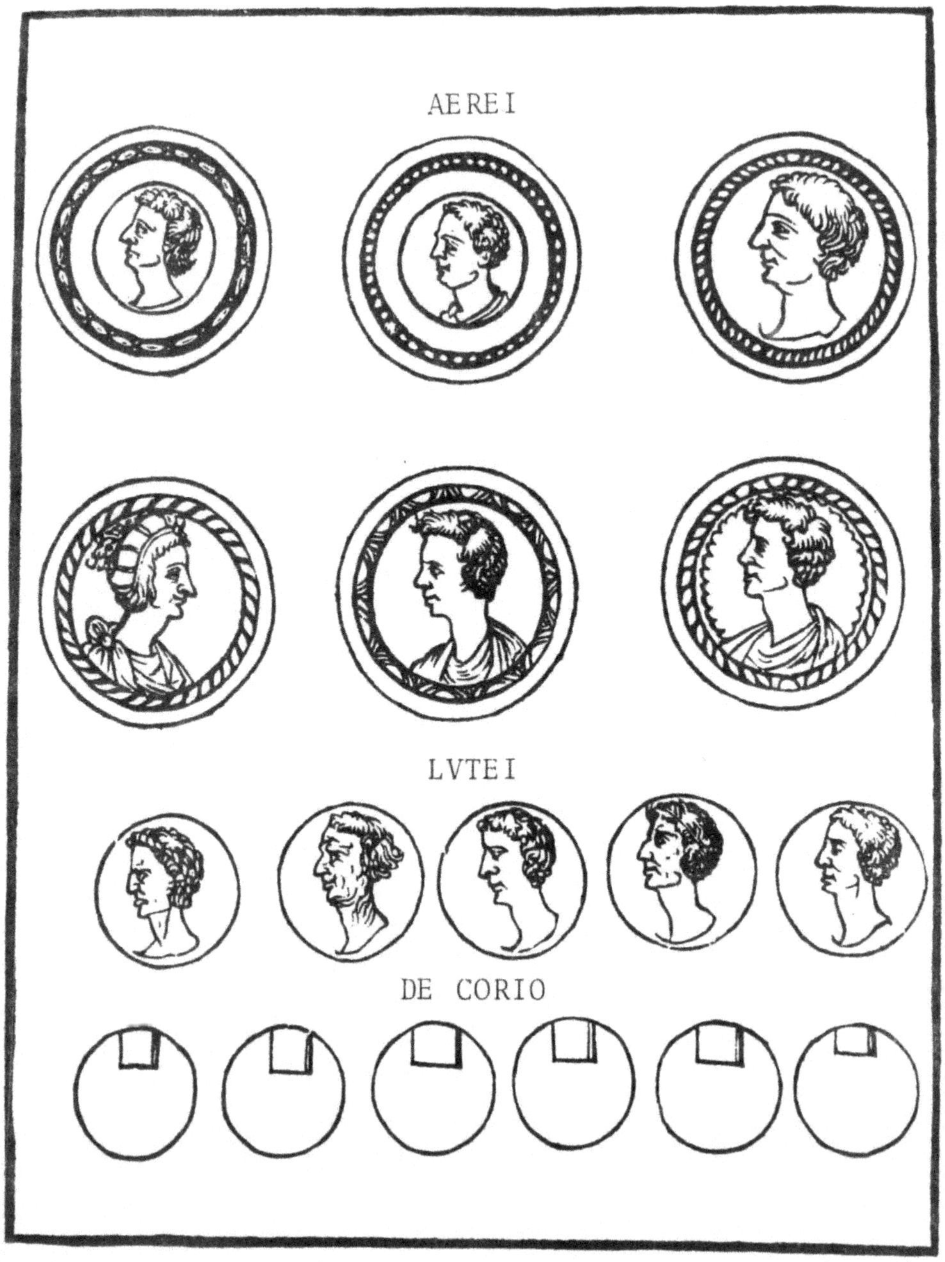

## i de inhibenda largitate

1 bellicam laudem et gloriam triumphorum utilitas semper *intueatur
aerarii, ne profusa largitio semina magis excitet proeliorum; quae
si prouidentia maiestatis reprimatur, non amplius bellorum florebit
2 improbitas sed collatorum potius defecta subsidia recreantur. quod
si largitio immoderata seruanda profuderit, opum delectabilitas
3 nullo uetustatis more poterit iam necessitatibus subuenire. quam
ob rem patrum nobis est paulisper prouidentia referenda quam in
rebus egenis habuerint, uel quae nunc **ita* in tanta facultatum
4 licentia uideatur. rectores superioris uitae non otiosis opibus
sed condendis potius moenibus laetabantur, in quorum decorem uni-
5 uersam auri argentique materiam conferebant. aeris quoque copiam
6 in simulacris propriis ad uirtutis suae testimonia figurabant. ut
uero emendi et uendendi utilitas * ***** et ut facultas regiae
largitati suppeteret, in aeris usum excultam politius terram et
igne solidatam, certis quoque expressionibus figuratam, auro argen-
7 toque reposito usui habuerunt. sed posteriores fastidientes priscae
uilitatis inuentum formatos e coriis orbes auro modico signauerunt,
quibus regum munificentia et commutationum necessitas intactis col-
8 latoribus utebatur. sequentior uero aetas, aeris redundante materia
quam publicus iam moenium recusabat ornatus, diuturna ut prior monu-
menta meditata aes ualidum ipso pondere pretiosius figurauit; cuius
species pro qualitate, ut diximus, ponderis diuturnior fuit, sed
potentia regalis pro licentia speciem suam tantum in auro argen-
toque signauit, quae pro reuerentia figurae nullis usibus profici-
9 ens ad honorem regium sacrata permansit. aeris autem materia,
quae iam pro copia uilior erat, ad dona militaria et uaria popu-
10 lorum commercia signabatur. sed ut uera fides dicta facilius pro-
sequatur, formas et species commodas atque pro temporibus diuersas
uariasque ueterum prouisiones exprimi colorum qualitatibus properaui

## ii ex quibus temporibus profusio uel auaritia coeperit

1 Constantini temporibus profusa largitio aurum pro aere, quod antea
magni pretii habebatur, uilibus commerciis assignauit; sed huius
2 auaritiae origo hinc creditur emanasse. cum enim antiquitus aurum
argentumque et lapidum pretiosorum magna uis in templis reposita ad
publicum peruenisset, cunctorum dandi habendique cupiditates accen-
3 dit. et cum aeris ipsius, quod regum, ut diximus, fuerat uultu sig-
natum, enormis iam et grauis erogatio uideretur, nihilo minus tamen
a caecitate quadam ex auro, quod pretiosius habetur, profusior ero-
4 gandi diligentia fuit. ex hac auri copia priuatae potentium repletae
domus in perniciem pauperum clariores effectae, tenuioribus uide-
5 licet uiolentia oppressis. sed afflicta paupertas in uarios scelerum
conatus accensa, nullam reuerentiam iuris aut pietatis affectum prae
6 oculis habens, uindictam suam malis artibus commendauit. nam saepe
grauissimis damnis affecit imperia populando agros, quietem latro-
ciniis persequendo, inflammando odia; et per gradus criminum fouit
tyrannos, quos ad gloriam uirtutis tuae produxit **inopia* magis quam
7 succendit audacia. erit igitur curae prudentiae tuae, optime
imperator, repressa largitate et collatori prospicere et in posterum

8 nominis tui gloriam propagare. denique paulisper felicium temporum
reuolue memoriam et antiquae paupertatis famosa regna considera
quae agros colere et abstinere opibus norant, qua haec honoris laude
9 per omne aeuum frugalitas incorrupta commendet. certe aurea nuncu-
pamus quae aurum penitus non habebant.

## iii de fraude et correctione monetae

1 inter damna rei publicae non ferenda, solidorum figura aliquantorum
fraudibus deprauata diuersa populos ratione sollicitat et regiae
maiestatis imaginem, dum per monetae culpam refutatur, imminuit;
ementis enim eundem solidum fraudulenta calliditas et uendentis
damnosa necessitas difficultatem quandam ipsis contractibus in-
2 tulerunt, ne rebus possit interesse simplicitas. ergo huic quoque
parti maiestatis uestrae est, ut in omnibus, adhibenda correctio:
ita ut opifices monetae redacti undique in unam insulam congregen-
tur, nummariis et solidorum usibus profuturi, a societate uidelicet
in perpetuum contiguae terrae prohibiti, ne commixtionis licentia
fraudibus opportuna integritatem publicae utilitatis obfuscet.
3 illic enim solitudine suffragante integra fides monetae praestabi-
4 tur, nec erit fraudi locus ubi nulla est mercis occasio. uerum
ut qualitas futurae discussionis appareat, formas et magnitudinem
tam aereae quam aureae figurationis pictura praenuntiante subieci.

FELIX INCHOATIO SACRAE DIVINAEQVE MONETAE

## iv de iudicum prauitate

1 ad haec igitur incommoda quae prouincias auaritiae artibus uexant,
accedit etiam iudicum execranda cupiditas collatorum utilitatibus
inimica: nam hi despecta reuerentia dignitatum uelut mercatores in
prouincias se missos existimant, eo grauiores quod ab his procedit
2 iniquitas unde debuit sperari medicina. et tamquam sua rebus suf-
ficere non possit iniquitas, exactores in profligandis rebus huius
modi dirigit unus quisque qui diuersis rapinarum artibus collatorum
uires exhauriant; uidelicet quasi parum notabiles haberentur si
3 soli peccarent. quae enim ab his occasio fiscalium titulorum il-
4 libata est? quae conuentio sine praeda discessit? illis tironum
comparatio, equorum uel frumenti coemptio, expensa quoque moenibus
5 profutura sollemnia lucra sunt et uotiua direptio. quod si prouin-
cias casti et integritatis cupidi moderentur, nec fraudi usquam
relinquetur locus et res publica moribus ditata firmabitur.

## v de releuando militari sumptu

1 relatis, ut uisum est, rei publicae incommodis augustis prouisi-
onibus merito repellendis, ad enormia militum alimenta ratione
non incongrua prohibenda ueniamus, quorum causa totius tribu-
2 tariae functionis laborat illatio. sed ne tantae confusionis
molem occupatio augusta fastidiat, breuius tam diuturni erroris
3 exitum declarabo. militaris ordo stipendiis aliquot peractis
ubi ad quinque uel eo amplius annonarum emolumenta peruenerit,
ne haec diutius percipiens rem publicam grauet, honesta missione
donatus uacans sibi otio gaudeat absolutus; in cuius locum poste-
rior succedens totum numerum per certa temporum spatia expensis
4 grauissimis releuabit. quod si numerosior miles de sequentibus
scholis in decedentium locum uocatur, hic quoque pari liberali-
tate *otio donatus abscedat uel ad alium ordinem cui miles deest
5 locum suppleturus accedat. quae res non solum expensis grauidam
rem publicam recreabit sed etiam curas imperatoriae prouisionis
imminuet; animabit etiam plures ad militiam quos ab ea stipendi-
6 orum tarditas prohibebat. huius modi igitur prouisionis utilitas
in augmentum prouincialium habet ueteranos regiis donis opulentos
et ad colendos agros adhuc praeualentes agricolas: habitabunt
limites, arabunt quae dudum defenderant loca, et laborum desiderio
7 potiti, erunt ex milite collatores. uerum quia non numquam bel-
lorum ruina aut fastidio castrensium munerum deserta militia de
summa integritatis intercipit, tali remedio huius modi damna sup-
plenda sunt, scilicet ut centeni aut quinquageni iuniores extra
hos qui in matriculis continentur habeantur in promptu armis exer-
citi et minori [ut minori] utpote tirones stipendio subleuati, in
8 locum amissorum si res ita tulerit subrogandi. his ita prouisis
et integritas secura manebit exercitus et damnis non deerunt matura
et parata in primis subsidia.

## vi de bellicis machinis

1 in primis sciendum est quod imperium romanum circumlatrantium ubi-
que nationum perstringat insania et omne latus limitum tecta natu-
2 ralibus locis appetat dolosa barbaries. nam plerumque memoratae
gentes aut siluis teguntur aut extolluntur montibus aut uindican-
tur pruinis; non nullae uagae solitudinibus ac sole nimio prote-
3 guntur. sunt quae paludibus fluminibusque defensae nec inueniri
facile queunt et tamen quietem pacis lacerant inopinatis incursibus.
4 ergo huius modi nationes, quae aut talibus subsidiis aut ciuitatum
castellorumque moenibus defenduntur, diuersis et nouis armorum sunt
5 machinis prosequendae. uerum ne qua difficultas in excitandis
armorum generibus oriatur, imaginem tormentorum nihil a uero distan-
tem coloribus adumbratam orationi subieci, ut sit facilis imitandi
confectio.

BALLISTA QVADRIROTIS

## vii expositio ballistae quadrirotis

1 exemplum ballistae, cuius fabricam ante oculos positam subtilis
2 pictura testatur. subiecta namque rotarum quattuor facilitas,
duobus subiunctis et armatis equis, ad usus hanc bellicos trahit;
cuius tanta est utilitas pro artis industria ut omni latere in
hostem sagittas impellat, sagittarii libertatem et manus imitata.
3 habet foramina per quattuor partes, quibus pro commoditate rerum
circumducta et flexa, facillime ad omnes impetus parata consistat.
4 quae quidem a fronte cochleae machina et deponitur celerius et
5 erigitur subleuata. sed huius temo in quamuis partem necessitas
6 uocet cita et facili conuersione deflexus erigitur. sciendum est

autem quod hoc ballistae genus duorum opera uirorum sagittas ex se non ut aliae funibus sed radiis intorta iaculatur.

CLIPEOCENTRVS TICHODIFRVS

viii expositio tichodifri

1 tichodifrus quod est genus machinae ex rei suae commoditate, graeca appellatione, uocabuli sumpsit exordium, eo quod per hunc facilior in murum paretur ascensus, ante ballistae semper ducendus incessum quo protectior eadem ballista operetur; erit ergo huius quoque com-
2 positio uel fabrica utili et commoda inuentione praeparanda. ergo hic idem tichodifrus non altior sed humilior fabricatur ut intra se possit latenter incedentium celare uestigia: qui, duabus superimpositis cratibus fixoriisque confixis, intra se tectos ab omni incursione defendit, duabus sane rotis ad promouendam machinam latenter
3 impulsus. cuius axium extremitates et frons nec non et superior latitudo fuscinis et lanceis armatur diligenter aptatis, ne aggrediendi cuiquam per uacantia defensore loca - id est superiorem partem - tribuatur facultas.

ix expositio clipeocentri

1 parma, hoc est modicus clipeus, fixoriis minutis ad soliditatem sui diligenter munitus, uarie habetur idoneus; interdum enim operit gestientem, non numquam in fronte tichodifri oppositus uel in temone suspensus munit inferius latitantes ab incursu comminus uenientum.

PLVMBATA MAMILLATA PLVMBATA [ET] TRIBOLATA

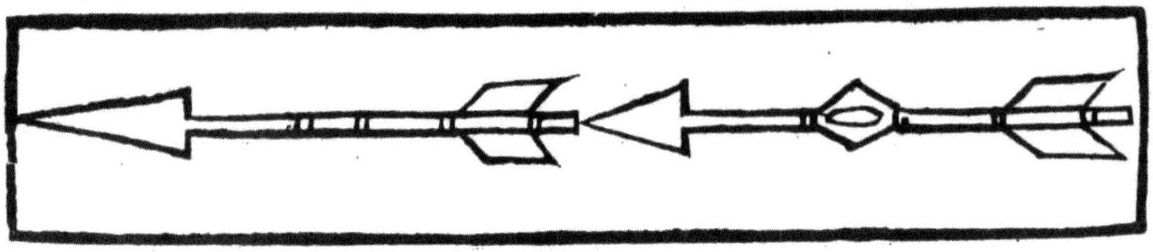

x expositio plumbatae [et] tribolatae

1 hoc iaculi genus, quod in modum sagittae pennis uidetur instructum,
non arcus neque ballistae pulsu consueuit emitti, sed manus impetu
et uiribus elisum in hostem comminus uadit: quod excipienti gemina
ratione uidetur afferre perniciem; aut enim perimit infixum aut
sine uulnere dilapsum et in terram positum plantae calcantis in-
figitur, eo quod in partem quamlibet si fuerit conuersum, ex latere
2 aculeo ex se eminente inferat uulnus. fit autem ex ligno in modum
sagittae facto, cui ferrum subtiliter in formam uenabuli aptatum
infigitur, fistula eiusdem ferri parumper extensa, supra quam,
modico interiecto spatio, plumbo adhaerentes aculei uelut tribuli
3 emergunt. in summa autem parte eiusdem iaculi affiguntur pennae
celeritatis causa, tanto uidelicet super easdem pennas relicto
spatio quantum digiti potuerint tenentis amplecti.

xi expositio plumbatae mamillatae

1 bene extensa et directa uirga accipiet in extremitate sui rotundum et in acumen deductum ferrum, similibus locis **ac* in tribulata plumbo et pennis adhaerentibus, ut plumbi pondere et pennarum celeritate adiuta rotunditas teli facilius clipeos aduersarii et similiter obstantia ualeat penetrare.

CVRRODREPANVS

xii expositio currodrepani

1 huius modi pugnacis uehiculi genus, quod armis praeter morem ui-
2 detur instructum, repperit parthicae pugnae necessitas. sed hoc
singulis bene munitis inuecti equis duo uiri uestitu et armis *e
ferro diligenter muniti citato cursu in pugnam rapiunt; cuius
posterior supra currum pars cultris in ordinem exstantibus com-
munitur, uidelicet ne facilis a tergo cuiquam praebeatur ascensus.
3 falces uero acutissimae axibus eiusdem currus aptantur, in late-
ribus suis ansulas habentes, quibus innexi funes pro arbitrio du-
orum equitum laxati quidem explicant, repressi autem erigunt falces.
4 qualia uero huius modi machinae funera hostibus immittant uel quas
turbatis ordinibus strages efficiant, dicent melius qui usu bella
cognoscunt.

CVRRODREPANVS SINGVLARIS

xiii expositio currodrepani singularis

1 in hoc curriculo, quod singularis equi tractu discurrit quodque unius hominis arte regitur et uirtute protegitur, eadem est quae in priori curru armorum facies et ad repellenda tela et ad inferenda pericula, nec distat a superiore quicquam machina nisi quod superior currus pro magnitudine sui duobus equis tractus, a duobus etiam bellatoribus gubernatur.

CVRRODREPANVS CLIPEATVS

xiv expositio currodrepani clipeati

1 machinae huius admiratio habet nouitatis aliquid; est enim a
superioribus curribus quadam parte dissimilis, hic enim posterior
uehiculi pars uerberibus spontaneis ad incitandos equos et clipeis
acuto ferro circumdatis uelut in propugnaculo positis communitur.
2 et est priori quidem currui hac ratione dissimilis, quod hic non
ut ibidem duos equos duorum etiam hominum sed unius tantum sagaci-
3 tas regit. a secundo uero curru hac diuersitate secernitur, quod
in illo unus equus et pondus bellatoris subit †dum ipsum equitem
4 portat et† occulto temone iugum trahente minus amplectitur. hic
etiam licet conuertere laxatis funibus explicatas falces et ad
5 perniciem hostium longius ab axe porrectas. funis uero, quo laxato
aut explicantur falces aut represso super eosdem axes eriguntur,
ligatur circulo haerenti catafracto in posterioris lateris parte
exteriore, per quod uestiti equi ad omnes telorum muniuntur incursus.

THORACOMACHVS

xv expositio thoracomachi

1 inter omnia quae ad usum bellicum prouida posteritatis cogitauit
antiquitas, thoracomachum quoque mira utilitate ad leuamen corporis
2 armorum ponderi et asperitati subiecit. hoc enim uestimenti genus,
quod de coactili ad mensuram et tutelam pectoris humani conficitur,
de mollibus lanis timoris sollicitudo sollertia magistra composuit
ut hoc inducto primum lorica uel cliuanus aut his similia fragili-
tatem corporis ponderis asperitate non laederent, membra quoque
uestientis inter armorum hicmisque discrimen tali solacio adiuta
3 labori sufficiant. sane ne idem thoracomachus pluuiis uerberatus
ingrauescente pondere adficiat uestientem, de libycis bene confec-
tis pellibus ad instar eiusdem thoracomachi faciem conueniet super-
4 inducere. hoc igitur, ut diximus, thoracomacho inducto (qui graeca
appellatione ex tuitione corporis nomen assumpsit), soccis etiam
(hoc est calciamentis) et ferratis ocreis inductis, superposita

galea et scuto uel gladio lateri aptato, arreptis lanceis, in plenum pedestrem subiturus pugnam miles armabitur.

ASCOGEFYRVS

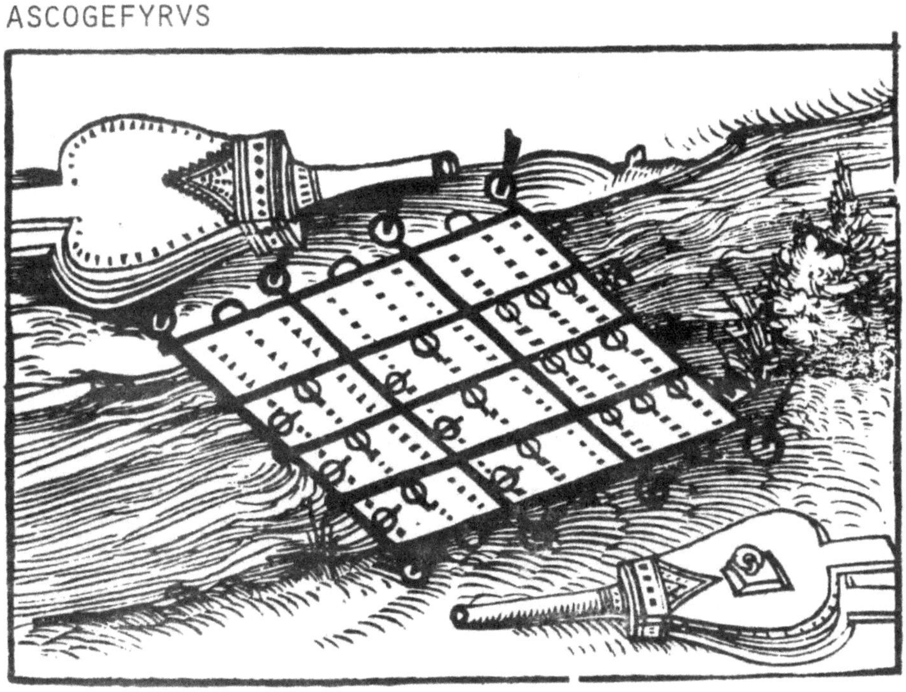

xvi expositio ascogefyri

1 ne interdum necessario itineri exercitus fluuiorum (quod plerum-
que euenit) occursus impediat, remedium ad hanc rem et compendio
facile et usu praecipuum repperit ingeniosa necessitas, quod tali
2 ratione componitur. uitulinis pellibus arabica arte mollitis -
est enim apud eos praecipua confectionis cura propter aquam de
puteis follibus hauriendam - his igitur ut dictum est, diligenter
sutis, fiunt utres in magnitudinem trium et semis pedum, ita ut
cum idem utres spiritu inflati tumuerint non extollant uterum sed
aequalitate quadam plenitudo ipsa utrium spatia plana perficiat,
ex quorum lateribus loris subter adnexis inuicem colligantur; de-
super autem una parte circulis exstantibus ex altera immittuntur
uncini: et ita in formam pontis adsociata partibus explicatur in-
3 tegritas. sed hoc idem opus obliquo fluuio propter impetum meatus
facilius usque ad alteram explicabitur ripam; quod, fixis in utra-
que ripa ferreis palis, et funibus ualidis in medio quidem sub
ipsis utribus propter incedentium sustinendum pondus, in margini-
bus autem firmitatis gratia desuper extensis, transeundi per flu-
uium nouo quodam et peregrino itineris apparatu intra breue tempo-
4 ris spatium praebebit liberam facultatem. admonendi praeterea
sumus quod super utrium compaginationem cilicia sunt incedentium
substernenda uestigiis ne lubrica pellium confectio insistendi
5 deneget firmitatem. in utraque tamen ripa erunt manuballistae
dispositae ne hostilis manus pontem operantibus impedimento con-
sistat.

LIBVRNA

xvii expositio liburnae

1 liburnam naualibus idoneam bellis, quam pro magnitudine sui
uirorum exerceri manibus quodam modo imbecillitas humana pro-
hibebat, quocumque utilitas uocet ad facilitatem cursus ingenii
2 ope subnixa animalium uirtus impellit. in cuius alueo uel capa-
citate bini boues machinis adiuncti adhaerentes rotas nauis late-
ribus uoluunt, quarum supra ambitum uel rotunditatem exstantes
radii, currentibus iisdem rotis, in modum remorum aquam conatibus
elidentes miro quodam artis effectu operantur, impetu parturiente
3 discursum. haec eadem tamen liburna pro mole sui proque machinis
in semet operantibus tanto uirium fremitu pugnam capescit ut omnes
aduersarias liburnas comminus uenientes facili attritu comminuat.

BALLISTA FVLMINALIS

## xviii expositio ballistae fulminalis

1 huius modi ballistae genus murali defensioni necessarium supra
ceteras impetu et uiribus praeualere usu compertum est: arcu
etenim ferreo supra canalem quo sagitta exprimitur erecto, uali-
dus nerui funis ferreo unco tractus eandem sagittam magnis uiri-
2 bus in hostem dimissus impellit. hunc tamen funem non manibus
neque uiribus militum trahi fabricae ipsius magnitudo permittit,
sed retro duabus rotis uiri singuli radiorum nisibus adnitentes
funem retrorsum tendunt, pro difficultate rei uiribus machinis
3 adquisitis. ballistam tamen ipsam ad dirigenda seu altius seu
humilius tela cochleae machina prout uocet utilitas nunc erigit
4 nunc deponit. hoc tamen mirae uirtutis argumentum: tot rerum
diuersitate connexum unius tantum otiosi (ut ita dicam) hominis
ad offerendam tantum modo impulsioni sagittam opera gubernat;
uidelicet ne, si hominum turba huius ministerio inseruiret, minu-
5 eretur artis inuentio. ex hac igitur ballista tot et tantis in-
genii artibus communita expressum telum in tantum longius uadit
ut etiam Danubii, famosi pro magnitudine fluminis, latitudinem
6 ualeat penetrare. fulminalis etiam nuncupata, appellatione sua
uirium testatur effectum.

7 his igitur, inuicte imperator, machinarum quoque inuentionibus
communitam potentiam inuicti duplicabis exercitus, hostium incur-
sibus non ui solum et uiribus obuiando sed etiam artis ingenio,
maxime cum animi sagacitate per omnia elementa machinas reppereris
8 ualituras. nam siue terra sint subeunda discrimina contra despe-
rati hostis audaciam, falcatis curribus per aperta camporum fugi-
entium agmina persequeris; siue murum subire temptauerit occulta
temeritas, experietur ballistae fidem ante quam ueniat; aut si
uictricem exercitus cursum fluuiorum uspiam meatus impediet, por-
9 tatilis pontis peculiaris succurret inuentio. quod si nauali bello
terras fugiens maria hostis obsideat, nouo celeritatis ingenio ter-
restri quodam modo ritu rotis et bubus subacta fluctibus liburna
10 transcurrens restituet sine mora uictoriam. quis enim huius uiribus
resistet, quae et terrestris uehiculi nititur firmitate et habet
11 de nauigii facilitate remedium? erit praeterea fulminalis ballista
felicium limitum custos, quae quanto minus ad armandam se uirorum
ministerium quaerit, tanto est ingenii uirtute praestantior; quic-
quid enim opis in ea numerosior manus afferre potuisset, id sibi
artis ingenio dotata largitur.

## xix de bellico apparatu

1 cum bellicus apparatus prouocet certamina, ne turba militum con-
fusis ordinibus uirtus compressa minuatur conueniet ducem spatia
ternis agminibus ordinare, ita ut singula quaeque agmina duobus
tantum uirorum milibus instruantur, ne compressa populositas, in
perniciem propriam auxiliis abundans, suis copiis opprimatur.
2 conueniet tamen per glaciales euntem militem plagas et thoracomacho
muniri et reliquis ad tutelam pertinentibus rebus prout membrorum
poscit utilitas armari, ut et frigoribus sufficiat et telis possit
occurrere; minoribus sane clipeis assumptis, ne sit (quod usu
plerumque euenit) pro armorum latitudine siluarum densitas inac-
3 cessa, et amittatur armorum enormitate refugium. ad omnia tamen
repentina subsidia militarem uiam leuibus portatum uehiculis asco-
gefyri comitetur auxilium, quo facilius et latior fluminum superetur
occursus et effugiendi trans fluuium, si res ita tulerit, facili-
4 tas offeratur. Persarum sane gens, cui praeter ceteras nationes et
dolus cordi est et corpori suppetit uirtus, quadratis est agminibus
et maiori bellorum apparatu superanda; ubi etiam machinarum est
omnis quam rettulimus industria proponenda, quae et uiribus resis-
5 tat et dolos callidae gentis refutet. in huius modi igitur con-
gressibus tichodifrus duorum gestatus ministerio uirorum protegat
6 ballistam multum bellicis subsidiis profuturam. sub ipsius ergo
tichodifri tegmine latentes scutati usque ad uiciniam pugnaturi
hostis adueniant, qui post detecti protegendi sunt ab omni telorum
nube, eiusdem tichodifri tabulatis, ut meminimus, acutis ferreis
7 communitis. cum uero inclinatum hostium pondus fugae praesidia
meditetur, falcatis curribus exercenda uictoria est quibus fugi-
8 entium poplites incidantur. ubi tamen angustiora loca agmen nequa-
quam explicari permiserint, coartatam aciem longus machinarum siue
tormentorum ordo circumdet, ne constrictum latus liberior hostis
9 infestet. erit tamen utilior cura si duplex animalium numerus ad
subuectionem tormentorum ducatur, ut sit copia in dubiis rebus uel
lassitudini uel casibus subuenire.

## xx de limitum munitionibus

1 est praeterea inter commoda rei publicae utilis limitum cura ambientium ubique latus imperii, quorum tutelae assidua melius castella prospicient, ita ut millenis interiecta passibus stabili muro et firmissimis turribus erigantur: quas quidem munitiones possessorum distributa sollicitudo sine publico sumptu constituat, uigiliis sane in his et agrariis exercendis, ut prouinciarum quies circumdata quodam praesidii cingulo inlaesa requiescat.

## xxi de legum uel iuris confusione purganda

1 diuina prouidentia, sacratissime imperator, domi forisque rei publicae praesidiis comparatis, restat unum de tua serenitate remedium ad ciuilium curarum medicinam, ut confusas legum contrariasque sententias, improbitatis reiecto litigio, iudicio augustae dignationis illumines; quid enim sic ab honestate consistit alienum quam ibidem studia exerceri certandi ubi, iustitia profitente, discernuntur merita singulorum?

# 2 TRANSLATION

## ON MILITARY MATTERS

### *Preface*

1 Most sacred emperors: advice upon the welfare of your empire, ever
prosperous under heavenly inspiration, should be offered on appro-
priate occasions, so that your glorious purposes may be confirmed
2 by glorious achievements. Consequently, I have written in this little
work, to the best of my ability, a chapter on the system of public
donations - not that this is a sufficient treatment of so enormous
a system, but so that from this preliminary evidence of my moderate
capacity the reliability of my expedients in what is to follow may
3 be displayed. And therefore - so that trust placed in what I promise
may not bear heavily on me hereafter, if it should appear that I
have deceived you - I request for myself, in consciousness that I
speak the truth, a punishment in place of a reward, if what I fore-
cast should turn out against my expectations; nor indeed should any
praise or reward attend me when it is quite enough, in the present
4 case, to escape your wrath at my temerity. But it will be admissible
that the head of state should find out from a private individual
what needs to be done, since the best solution of problems sometimes
5 escapes his inquiry. And so, from time to time, those who have been
shown to view anything correctly must be called in; for as an ex-
cellent orator says, 'the greater part of mankind has immediate
recourse to the man whom nature has endowed with the faculty of
6 thought'. In this, one must consider what a man thinks rather than
says; for everyone agrees that neither the loftiest birth nor the
abundance of wealth or the influence proceeding from public positions
or eloquence acquired in study has led to improvements in the arts
(in which the invention of military equipment has its place also),
but intellectual capacity alone, which is the mother of all excel-
7 lences, and depends on a happy accident of nature. And this we see
happen without respect of persons: for although the barbarian peoples
are neither potent in eloquence or ennobled by official rank, they
are nevertheless known to be not in the least strangers to the dis-
8 covery of things, when nature lends her aid. Wherefore, most merciful
emperors, you who in perpetual happiness love the glory of high
renown, who encourage in your sons the sentiments due to the name
of 'Roman', condescend to look upon what a timely providence from on
9 high has impressed on my understanding. Thus, as all rejoice in the
happiness of your reign, whether they engage in Your Clemencies'
service or are content with the peaceful life of a private citizen,
or are farmers of the land or merchants handling the profits of
commerce, you will pursue, guided by this work, the best interests
of each; the nature of which interests the following disquisition
will display, at the most appropriate points, in its various sections.
10 For I shall describe how, with a reduction of the taxes by one-half,
the farmer in the provinces may be restored to his traditional pro-
sperity; how also, with the abolition of abuses in the system of
taxation, the frontier-dweller may do honour to the lonely stretches
of the imperial boundaries in safety, after the erection of forti-
fied defences; and in what way the supply of gold and silver may be

doubled without hardship to the tax-payers; even by what means
the army may rejoice under a shower of honours beyond those which
11 are customarily granted. To this I have thought it necessary to
add provisions for the requirements of warfare by land and sea in
order to secure victories; amongst which, for the purpose of
lessening distaste, I will describe a small number of mechanical
12 inventions. Thus, I shall reveal that an exceptionally fast type
of warship, under the control of an ingenious device, can have the
advantage over ten ordinary ships, with the result that they can
13 be sunk by it without its needing a large crew. Whereas for engage-
ments on land a cunning arrangement has been devised such that a
horse, whether he is galloping forward to break through the enemy
line or whether he is bearing down upon fugitives, is equipped with
a contrivance such that he whips himself on and wreaks enormous
14 destruction amongst the enemy without anyone to control him. Again,
where difficult river-crossings are concerned, a design has been
thought out for a new type of bridge which is not at all heavy to
carry: a very few men, or about fifty pack-animals, will be able
to transport this bridge, which is essential for use across rivers
15 or marshy territory. In what I have described, therefore - now
that I have, I think, sketched what is to come - I bring you (allow
me to say so) a great gift by heaven's granting, stating as a fact
that by Your Pieties' foresight the power of the army, and the en-
tire empire, can be sustained by the measures which I have just
mentioned: these are, of course, already known to Your Clemencies'
most intimate advisers, who have a great number of other cares
16 which are foreign to me. But because many things escape their
attention amidst the pressure of business, it has been my urgent
desire - prompted as I am by retirement, and being not entirely
ignorant of what should be done for the best in these matters -
to bring together what I have collected from all quarters which
17 may be of use to Your Felicities. If, however, because of the re-
quirements of the subject, my representation produces something
rather outspoken, I believe that this will be favoured with pardon,
since in the fulfilment of my undertaking indulgence will be
accorded me in the interest of freedom of speculation.

[picture: THE VARIOUS SUITABLY-AUTHORIZED COINS OF THE ANCIENTS (BRONZE; CLAY; MADE OF LEATHER)]

I *Strict control of public donations*

1 The treasury, in its own interests, must always keep a watchful eye
on military success and the pomp of triumphs, to prevent the extra-
vagant donation of money from stirring up the seeds of conflict
to an even greater extent: if this is put down by Your imperial
Majesty's foresight, the evil of war will no longer flourish, but
instead the collapsing resources of the taxpayers will be refreshed.
2 Whereas if the uncontrolled distribution of money wastes what ought
to be preserved, the delights of wealth will not be able in any way
3 to supply our needs as they did in antiquity. Consequently I must
speak briefly on the subject of the foresight (or what would seem
such, in our present profligacy of resources) exercised by our
4 ancestors in an age of poverty. The rulers of earlier times took
pride not in useless treasure, but in the construction of buildings
instead, to beautify which they devoted their entire stock of gold
5 and silver. Their supply of bronze, too, they struck in represen-
tation of their own likenesses, as testimonies to their excellence.
6 But so that the convenience of buying and selling ....... * and so
that the means of royal bounty might be supplied, they stored up
their gold and silver, and, to take the place of bronze, they made
use of clay artfully treated, hardened by fire and impressed with
7 definite symbols. Those who came after, however, despising this con-
trivance of the humble days of old, made discs of leather and
attached to these a small piece of gold: with these the munificence
of the kings, and the requirements of trade, could be exercised
8 without harm to the taxpayers. But the following age - in which
there was a vast supply of bronze which could no longer be absorbed
in the decoration of public buildings -, intending, like the pre-
vious one, a monument for all time, struck heavy bronze coins, all
the more valuable precisely because of their weight: as a result of
their weight, which I have just noted, the types were very long-
lasting; but these powerful kings impressed their own image as they
wished only on gold and silver, and this image remained - because
of the sanctity of what was represented - a tribute to the kingly
9 state, useless for all practical purposes. Bronze, however, which
was now very cheap because of its abundance, was struck for purposes
10 of military donatives and the various civil transactions. But so
that complete conviction may the more readily attend on what I have
said, I have taken pains to have the shapes and designs which were
deemed suitable (and which differ according to their periods), and
the changing arrangements of antiquity, illustrated in their various
colours.

**Lacuna* (*see critical note*): *supply perhaps* might continue to benefit the people,

## II *The period at which extravagance and greed began*

1 In the age of Constantine extravagant public spending prescribed
gold for petty transactions instead of bronze, which previously
was highly valued; and the origin of this display of greed is
2 believed to have sprung from the following cause. When the gold
and silver and the enormous quantity of precious stones which had
been hoarded up many years ago in the temples reached the public,
3 it inflamed everyone's desires to spend and possess. And although
the spending of bronze itself (which, as I have said, had been
stamped with the portraits of kings) could already be seen to be
excessive and difficult to control, nevertheless, as a result of
some blindness or other, there came about an increasingly wide-
spread insistence on spending gold, which is understood to be
4 more valuable. The private mansions of the powerful, filled from
this flood of gold, were turned into more signal instruments for
the oppression of the poor; the less-well-off, obviously, being
5 held down by brutal means. But the suffering poor, driven to
attempt various acts of wickedness, and having before their eyes
no respect for the law nor feelings of loyalty, entrusted their
6 revenge to the criminal arts. For frequently they inflicted
extremely serious damage on the empire by laying the countryside
waste, by persistently disturbing the peace with acts of brigand-
age, by stirring up hatreds; and they encouraged along the stages
of their criminal career the tyrants whom desperation brought
forth (rather than emulation inspired) against Your virtuous
7 Splendour. Your Prudence, most excellent emperor, will therefore
take steps to hold down public spending, and thereby both to look
to the interests of the taxpayer and to extend to ages to come
8 the splendour of your name. Last of all, think over briefly what
is known of the happier ages of mankind, and remember the celebra-
ted kingdoms of the ancient times of poverty, which had learned
how to cultivate the land and not covet riches, remember with
what honour and praise their untainted frugality gives them a
9 name for all time. Certainly we speak of those ages as 'golden'
which had absolutely no gold at all.

## III *Fraud, and the reform of the mint*

1 Among the injuries to the state which cannot be tolerated, the
condition of the solidi, which has been debased through the frauds
of certain persons, troubles the people in different ways, and
cheapens the representation of imperial majesty when it is rejec-
ted through the fault of the mint; for the deceitful cunning of
the buyer and the miserable plight of the seller of one and the
same solidus have imported something of a problem into the con-
tracts concerned, so that a straightforward transaction in such
2 cases is impossible. And so Your Majesty's correction must be
employed, as in all other matters, so in this also, on the plan
that the workers of the mint should be brought in from all sides
and placed all together on one island, to occupy themselves with
the business of the lower denominations and the solidi, cut off,

that is, from contact with the neighbouring mainland for ever, so that freedom of passing to and fro, which provides opportunities
3 for fraud, may not blacken the integrity of a public service. For there, given the advantage of isolation, the perfect good faith of the mint will be guaranteed, and there will be no place for fraud
4 where there is no chance of profit. Now so that the types of a future issue may be seen, I have set out below in an informative picture the styles and size of both the bronze and the gold specimens.

[picture: A FAVOURABLE NEW START FOR THE SACRED AND DIVINE CURRENCY]

## IV *Corruption among the provincial governors*

1 On top of these inconveniences, then, which vex the provinces with the refinements of avarice, there comes also the abominable greed of the provincial governors, which is prejudicial to the interests of the taxpayers: for, caring nothing for the dignity of their office, they regard themselves as being sent into the provinces like merchants, all the more objectionable in that iniquity proceeds from the very men from whom a remedy was to be
2 expected. And as though their own iniquity were not enough for this state of affairs, each one of them, to further the work of destruction, sends out tax-collectors to drain the resources of the taxpayers by various refinements of extortion - evidently because they themselves would be thought totally unremarkable if
3 they were the only ones to do wrong. In fact, what opportunity of gain offered by the tax-returns have these people let slip? What proceeding for unpaid taxes has ever gone by without their
4 extracting their plunder? To these men the enlistment of recruits, the purchase of horses and corn, even the sums allocated for defence-walls, are all time-honoured sources of profit and the
5 eagerly-awaited opportunity for robbery. Whereas if honourable men, lovers of honesty, were in charge of the provinces, there will be no place for fraud anywhere, and the state will be strengthened by a rich accession of moral qualities.

## V *Reduction of military expenditure*

1 Having now, I think, reviewed the troubles of the state which must be abolished - and rightly so - by imperial measures of reform, let me turn to the vast expenditure on the army, which must be cut down in a not dissimilar fashion: because of this expenditure, the
2 whole system of tax-collection is in trouble. But so that Your Majesty, pressed as you are with business, may not grow weary at

the enormous mass of confusion here, I will set out very briefly
3 the solution of this inveterate problem. When a serviceman has
served for a certain number of years and has reached the point at
which he has received the rewards of five donatives or more, he
should be given an honourable discharge and enjoy his retirement
freed from his commitment, his time being all his own, which will
prevent his being a burden on the state by receiving donatives
year after year; the man next below in seniority will take his
place, and so at regular intervals of time will relieve the
4 battalion as a whole from tremendous expense. If, however, there
are too many soldiers due for promotion from the lower companies
into the place of those who are retiring, this surplus too can be
granted its retirement and leave the service, or it can be drafted
to another unit which is short of men, to fill up their place.
5 This plan will not only give new strength to the state, which is
burdened with expenses, but it will also reduce the cares of its
imperial director; it will, moreover, encourage many to join the
6 service whom the slowness of promotion deterred from it. And then
the utility of an arrangement of this kind is that it turns out,
to the benefit of those in the provinces, veterans enriched with
imperial grants and farmers still powerful of limb to till the
fields: they will inhabit the frontier-territory, they will plough
the lands which they but lately defended, and, having gained the
longed-for reward of their labours, from being soldiers they will
7 become taxpayers. But because the service is frequently depleted
by military disaster or boredom with the duties of camp life, so
cutting the full strength, losses due to these causes are to be
made good by a remedy of this kind, namely, that divisions of a
hundred or fifty younger men, in addition to those who are listed
on the pay-rolls, should be kept ready for action, trained in the
use of weapons, and paid (as being recruits) at a lower rate, to be
drafted into the place of those who have been lost if events should
8 make this necessary. Given provisions of this kind, the full
strength of the army will be maintained intact, and a supply of
excellently-trained recruits will be at hand in good time against
losses.

## VI *Military machines*

1 First of all it must be recognized that frenzied native tribes,
yelping everywhere around, hem the Roman empire in, and that
treacherous barbarians, protected by natural defences, menace
2 every stretch of our frontiers. For these peoples to whom I
refer are for the most part either hidden by forests or lifted
beyond our reach by mountains or kept from us by the snows;
some, nomadic, are protected by deserts and the blazing sun.
3 There are those who, defended by marshes and rivers, cannot even
be located easily, and yet they tear peace and quiet to shreds
4 by their unforeseen attacks. Tribes of this kind, therefore,
who are protected either by natural defences such as these or by
the walls of towns and towers, must be attacked with a variety
5 of novel armed devices. And so that no difficulty may arise in
constructing the various types of weapons, I have added beneath

my text a perfectly accurate picture of each machine, picked out in colours, so that it may be easy to build in imitation of it.

[picture: THE FOUR-WHEELED BALLISTA]

## VII *Description of the four-wheeled ballista*

1 A type of ballista, whose construction the detailed drawing sets
2 clearly before your eyes. Evidently, four smoothly-running wheels
move this around for battle purposes, two mail-clad horses being
yoked to it; and so versatile is it, thanks to its ingenious
design, that it can fire arrows in all directions against the
enemy, in imitation of the flexibility and dexterity of a bowman.
3 It has openings in its four sides, through which it is turned and
aimed as circumstances require, and so it can be got ready against
4 any attack with great ease. By means of a screw-thread device at
the front it is very quickly lowered, and as quickly lifted into
5 the raised position. Its pole, too, can be swung round by a
rapid and easy turning movement, and then raised, to face whatever
6 direction the need of the situation dictates. It must also be
pointed out that this type of ballista fires its arrows on being
wound up not by ropes, as others are, but by a windlass, which is
worked by two men.

[picture: THE *CHAR-MURAILLE*: THE SHIELD *A CLOUS*]

## VIII *Description of the* char-muraille

1 The *char-muraille* type of machine has taken the origin of its name
from a Greek designation derived from the nature of its function,
in that by means of it the scaling of a wall is made easier: it is
always to be brought up in front of a ballista moving into action,
so that this ballista may operate under better cover. The assembly
or construction of this machine too, then, is to be carried out in
2 accordance with a practical and proper design. The present *char-
muraille*, therefore, is built not high off the ground but low down,
so that it can conceal within itself all traces of its crew, who
advance in hiding; with two wicker screens placed above it, and its
nails hammered well home, it protects the men hidden inside it from
all attacks, being moved, invisibly, by two wheels intended to propel
3 the superstructure. The tips of its axles, and its front end, and its
upper surface too, are armed with carefully-fitted prongs and blades,

so that no-one may be given the opportunity of making an assault on the undefended area - that is, the upper part.

## IX *Description of the shield* à clous

1 A targe (that is, a small shield), carefully studded with tiny nails to strengthen it, is thought desirable for various reasons: sometimes, that is, it protects the man who carries it; sometimes, fastened on the front of the *char-muraille* or hanging from its pole, it guards the men hiding beneath from the attacks of opponents coming at close quarters.

[picture: THE 'CALTROP' JAVELIN; THE 'BREASTED' JAVELIN]

## X *Description of the 'caltrop' javelin*

1 This type of javelin, which can be seen to be equipped with flights like an arrow, has been intended to be hurled not by the impulse of a bow or a ballista; flung instead by the thrust and power of the hand, it travels at close range into the enemy: this can be seen to threaten death in one of two ways to the man who is wounded by it; for it either buries itself in his body and kills him, or it falls away without wounding him and lies on the ground and sticks in his foot when he treads on it, because, whatever way it is turned, it can inflict a wound from its edge
2 with the prong projecting from it. Now it is made of a piece of wood fashioned like an arrow, to which is carefully fixed an iron tip worked into the shape of a hunting-spear, with the tubular end of the blade extended a little way: above this, with a short gap left in between, spikes like caltrops project, soldered on
3 with lead. And at the lower end of the javelin are fixed flights to give speed, with enough space left above these flights, of course, for the fingers of the holder to be able to grasp.

## XI *Description of the 'breasted' javelin*

1 A shaft, nicely long and straight, will have fitted to one of its ends a piece of iron, round in section and tapering to a point, with a lead weight and flights attached at the same points as in the 'caltrop' type, so that the bulbous weapon, assisted by the weight of the lead and the swiftness of the flights, will be

powerful enough to penetrate very easily the enemy's shields and similar obstacles.

[picture: THE CHARIOT *A FAUX*]

## XII *Description of the chariot* à faux

1 The peculiar requirements of the Parthian war led to the invention of this type of fighting vehicle, which can be seen to be
2 provided with offensive weapons in a remarkable fashion. Two men, however, properly equipped with steel clothing and weapons, and each riding on a well-protected horse, drive this particular machine into battle at great speed: its rear end is equipped with knives projecting in a row above the chassis - obviously so that no-one may be afforded an easy chance of climbing on it
3 from the back. Furthermore, very sharp scythes are fitted to the axles of this chariot, with eyelets on their edges, in which are knotted ropes which, when released at the will of the two riders, let down the scythes, and when tightened, lift them up.
4 Those who know war through personal experience will tell better than I what destruction devices of this kind can wreak among the enemy, or what carnage they can produce among the shattered ranks.

[picture: THE CHARIOT *A FAUX* FOR ONE RIDER]

## XIII *Description of the chariot* à faux *for one rider*

1 In this chariot, which moves about under the pull of one horse and which is directed by the skill and defended by the courage of a single man, the arrangement of the offensive devices, both for repelling weapons and for thrusting home the attack, is the same as in the previous chariot: it has, in fact, no differences at all from the machine described above, except that the chariot described above is hauled (because of its great size) by two horses, and is also driven by two warriors.

[picture: THE CHARIOT *A FAUX* WITH SHIELDS]

## XIV *Description of the chariot* à faux *with shields*

1 This amazing machine incorporates a certain novelty, being in
one respect different from the chariots described above; for here
the rear end of the vehicle is equipped with automatic whips to
goad the horses on, and with shields, surrounded with sharp iron
2 points placed as on a battlement. It differs, too, from the
first chariot in this way, that here the skill not of two men,
3 as there, but of one man only directs the two horses. It is,
moreover, distinguished from the second chariot by this point
of difference, that there a single horse both takes the warrior's
weight in carrying the rider [? *a gloss*] ... , and surrounds ...
to a lesser extent, since a hidden pole pulls the yoke [*or:* and
the yoke surrounds ... to a lesser extent, since a hidden pole
4 pulls ...].* Here too one can change the position of the scythes,
which, when the ropes are slackened, are extended and stretch far
5 beyond the axle to slaughter the enemy. The rope, by means of
which the scythes are either extended (when it is released) or
lifted above the axles (when it is tightened), passes through a
ring attached, on the outer side of the rear flank, to the scale-
armour by which the horses are clad and protected against all
blows of weapons.

**The sense of this defective and perhaps interpolated passage may originally have been:* ... that there a single horse both takes the warrior's weight in carrying the rider and <draws the vehicle, being placed between two shafts; whereas this chariot> surrounds <its horses> to a lesser extent, since a hidden pole pulls the yoke (*so Hassall*): *but the Latin is beyond recovery* (*see critical note*).

[picture: THE *DEFENSEUR DE POITRINE*]

## XV *Description of the* défenseur de poitrine

1 The ancients, among the many things which, in their forethought
for posterity, they devised for use in war, prescribed also the
*défenseur de poitrine* to counteract the weight and friction of
2 armour: it is amazingly useful for protecting the body. This
type of garment is made of thick cloth to the measure and for
the protection of the upper part of the human frame; fearful
apprehension, guided by cleverness, devised it, so that, after
it has been put on first, the breastplate, or suit of armour, or
something similar, cannot injure the frail body by its roughness
and weight; and again, the limbs of the wearer, helped by this
means of relief, will be able to do their work amidst the diffi-
3 culties of warfare and cold weather. But in any case, so that
the *défenseur de poitrine* may not cause problems for the wearer

with its increasing weight when it is soaked with rain, it will be advisable to put over on top a covering garment made of nicely-treated Libyan hide in the shape of the *défenseur de*
4 *poitrine* itself. So, when the *défenseur de poitrine* has been put on as I have described - it derives its name through Greek nomenclature, from its protecting the body - and with pumps (footwear, that is) and steel greaves put on, a helmet on his head and a shield and sword slung by his side, snatching up his spears, the soldier will be completely armed, ready to go into an infantry engagement.

[picture: THE *PONT D'OUTRES*]

XVI *Description of the* pont d'outres

1 To prevent encounters with rivers from delaying (as very often happens) a march which it is sometimes necessary for the army to make, inventive necessity has discovered a remedy against this difficulty, at once economically simple and outstandingly
2 effective; it is made in the following manner. Calf-skins are softened by the Arabian process - because in that country there is a particularly well-developed technique of treatment, as a result of their need to lift water from wells in buckets - skins of the sort that I have mentioned, then, are stitched together carefully to form bladders three-and-a-half feet across, in such a way that, when these bladders are filled with air and swell up, they do not bulge outwards, but instead the inflation of the bladders produces flat surfaces by a kind of equal stretching: they are tied to each other, side by side, by means of thongs knotted underneath: on top, small loops project on one side, and hooks are slipped into these from the other; and thus the whole thing, put together in the form of a bridge, is built up complete
3 in its parts. But this structure will be more easily built out towards the other bank if the river strikes it obliquely, because of the push of the current; and then, with iron stakes fixed in both banks, and with strong ropes running both centrally under the bladders to take the weight of those walking on them and also along the top of the sides to give rigidity, the structure will offer an easy means of crossing a river in a short space of time
4 by means of a new and unfamiliar item of marching-equipment. In addition, we need to remember that hair mats have to be placed above the assembly of bladders where those walking across will tread, so that the greasy dressing of the skins does not prevent
5 a firm foothold being taken. Again, hand-ballistae will be stationed on both banks, so that an enemy detachment cannot get in the way of those working on the bridge.

[picture: THE WARSHIP]

## XVII *Description of the warship*

1 Animal power, directed by the resources of ingenuity, drives
with ease and swiftness, wherever utility summons it, a war-
ship suitable for naval combats, which, because of its enormous
size, human frailty as it were prevented from being operated by
2 the hands of men. In its hull, or hollow interior, oxen, yoked
in pairs to capstans, turn wheels attached to the sides of the
ship; paddles, projecting above the circumference or curved
surface of the wheels, beating the water with their strokes
like oar-blades as the wheels revolve, work with an amazing
3 and ingenious effect, their action producing rapid motion. This
warship, moreover, because of its own bulk and because of the
machinery working inside it, joins battle with such pounding
force that it easily wrecks and destroys all enemy warships
coming at close quarters.

[picture: THE 'THUNDERBOLT' BALLISTA]

## XVIII *Description of the 'thunderbolt' ballista*

1 This type of ballista, essential for the defence of fortifications,
has been found in practice to be superior to others in force and
power; this is because an iron bow is placed vertically above the
channel along which the arrow is propelled, and a strong rope of
elastic fibre is pulled back with an iron hook and fires this arrow
2 with enormous power into the enemy when it is released. The great
size of the construction does not allow this rope to be pulled back
by the power of soldiers' hands: instead, two men, pushing at the
spokes of two wheels placed at the rear, one man to each, stretch
the rope back, power proportionate to the difficulty of the opera-
3 tion being obtained by means of machinery. A screw-thread device,
too, now raises and now lowers the ballista itself, so that it may
4 fire its arrows higher or lower as occasion dictates. But now the
proof of its amazing and unique excellence: though it is composed
of so many different parts, it is controlled by the attention of
one man only, who is (so to speak) at his ease, for the purpose
simply of loading the arrow in front of the propulsion-mechanism;
this, you see, avoids impairing the basic purpose of the design,
as would happen if a large number of men performed this function.
5 A missile, then, fired from this ballista endowed with so many
great refinements of ingenuity, flies such a very long way that
it can even traverse the width of the Danube, famed for the breadth
6 of its stream. And the machine's designation - the 'thunderbolt' -

bears witness by its very name to the effectiveness of its power.
7 And so, invincible emperor, you will double the strength of your
invincible army if it is reinforced with these mechanical inven-
tions too, checking enemy invasions not merely by force and might
but also by scientific devices: particularly since you will have
discovered, in the keenness of your mind's perception, that the
8 machines will be of service in all physical conditions. For if
battles are to be fought out on land against the boldness of a
desperate enemy, you will pursue his fleeing forces across the
open plains with the scythed chariots; if a rash foe attempts
secretly to scale a wall, he will find out how trusty a thing the
ballista is before he gets anywhere near; or if streams of water
anywhere block the victorious advance of the army, my special
9 device of a portable bridge will lend its aid. And if the enemy
flees the land and invests the seas with naval conflict, the war-
ship, racing across the waves with its new contrivance for fast
travel, moved along by wheels and oxen, in a sense as though it
10 were on land, will give instant victory. For who will be able to
resist the power of this ship, which rests on the stability of a
land vehicle and offers the advantage of a ship's manoeuvrability?
11 In addition, too, the 'thunderbolt' ballista will guard your
fortunate frontiers: the less need this has of men's assistance
to put it into firing order, the more remarkable is it in the
unique excellence of its contrivance, in that it provides for
itself, being endowed with the contrivances of science, all the
resources that a larger crew might afford it.

## XIX *Military equipment*

1 When the display of military might provokes engagements, the
commander-in-chief will find it advisable to arrange breaks be-
tween every three columns of men, so that the fighting-power may
not be restricted and reduced by disorganization of the ranks
caused by too large a number of soldiers: given also that each
column should consist of only two thousand men, to prevent the
overwhelming by its own forces of a very large number of men all
packed together, with resources so abundant that they actually
2 inflict damage upon themselves. And then it will be necessary for
the soldier passing through ice-bound regions to be equipped with
the *défenseur de poitrine* and to be armed with whatever else is
needed to protect him, as the requirements of his body dictate,
so that he can both resist the cold and face missiles: small
shields will obviously be carried, so that dense woodland may not
be inaccessible because of the width of the equipment (which
often happens in practice), and so that the opportunity of con-
cealment may not be lost because of the huge size of the equipment.
3 For all emergency purposes the helpful *pont d'outres* should ac-
company the soldiers' march, transported on small vehicles, so
that wide rivers which are encountered may be crossed easily, and
also so that - if circumstances make this necessary - there may
4 be at hand an easy means of escaping across the water. As for the
Persians, who above all other nations are addicted to treachery
and possess great physical strength, they must be destroyed by

means of square formations and an enormous quantity of battle-
gear: here too the operation of all the machines which I have
described must be brought forward to resist the power and confute
5 the treacheries of this cunning race. In battles of this sort,
then, the *char-muraille*, carried along by the agency of two men,
can protect the ballista which will be an invaluable part of the
6 combat-equipment. Men carrying shields, hiding under the protec-
tion of this *char-muraille*, can come right up to the neighbour-
hood of the enemy when he is on the point of giving battle; and
then, when they come out into the open, they will be protected
from all the missiles raining down, since the platforms of the
*char-muraille* are defended (as I have described) by iron spikes.
7 And then, when the enemy force yields and is intent on seeking
safety in flight, the decisive blow will be administered by the
scythed chariots, by means of which the fleeing enemy's hamstrings
8 can be cut. Where, however, very restricted conditions prevent
the army from being fully deployed, a long line of machines or
throwing-engines should surround the condensed force, so that the
9 enemy may not be free to molest the packed flank. Further, it
will be a very useful plan if the number of animals brought along
to haul the throwing-engines is doubled, so that at critical
moments there is a reserve to replace those which are exhausted
or have been killed.

## XX *Defence of the frontiers*

1 A proper concern for the frontiers which surround the empire on all sides is also to the advantage of the state: an unbroken chain of forts will best assure the protection of these frontiers, on the plan that they should be built at intervals of one mile, with with a solid wall and very strong towers; the landowners would construct these defences on their individual responsibility, at no expense to the state, and of course watches and country patrols should be organized in them, so that the peace of the provinces, protected by a kind of belt of vigilance, may rest unharmed in quiet.

## XXI *Removal of confusion in the laws and the legal system*

1 Most sacred emperor, when measures for the defence of the state
at home and abroad have been furnished by Your divine Providence,
there is one remedy yet required of Your Serenity to heal our
domestic troubles: that by the power of judgement belonging to
Your imperial Condescension you should throw light upon the con-
fused and contradictory pronouncements of the laws, casting aside
2 the disputes of evilly-disposed men. For what is there more alien
from honourable conduct than that the passion for captious debate
should be given rein in the very place where, from the mouth of
justice, distinctions are made between the merits of individuals?

# 3 THE TRANSMISSION OF THE TEXT

## THE TRANSMISSION OF THE TEXT

1. Five principal witnesses to the text of the group of works which includes *de rebus bellicis* survive, all descended from a lost manuscript (to be described in due course) which belonged to the cathedral library at Speyer. These primary sources, with their accepted *sigla*, are:-

C, at Oxford, *oxoniensis canonicianus class. lat.* 378 (*de rebus bellicis* on ff. 67r-77v);

M, at Munich, *monacensis latinus* 10291 (*de rebus bellicis* on ff. 171-222);

P, at Paris, *parisinus latinus* 9661 (*de rebus bellicis* on ff. 53v-63r);

V, formerly at Vienna, *uindobonensis* 3103, now at Trento (*de rebus bellicis* on ff. 53r-63v);

and the typographic *editio princeps*, prepared by Siegmond Ghelen ('*Sigismundus Gelenius*'), and printed by Froben at Basel in 1552 (*de rebus bellicis* on sigs. q1r-r3v).

Other manuscripts which carry the treatise - *florentinus laurentianus* 913, *leidensis uossianus* 44, *matritensis* Q129, *monacensis latinus* 794 [imperfect], *parisini latinus* 5825E and *nouv.acq.*1424, *uaticani latinus* 3715 [imperfect] and *barberinus latinus* 157(809), *uindobonensis* 3102, and the *parmensis* seen by Panciroli,[1] are variously derived from C M P or V and make no independent contribution to the establishment of the text.[2] In addition, the equally derivative Copenhagen manuscript *gottorpiensis* 498 contains the pictures but not the text of the work.

The precise relationship of three of the primary sources to their archetype is ascertained from external testimony: that of the other two must be conjectured from internal evidence, and the reader is warned, once for all, that my conclusions in this respect, as in others, are more severe than those admitted by previous editors of the components of the Speyer corpus. Leaving discussion of these disputed cases (namely P and the *editio princeps*) until last, we may set out briefly the known facts concerning the manuscripts C M and V.

2. C was transcribed directly from Σ, the *spirensis*, in or about January 1436, for Pietro Donato, whose colophon states:

> Exemplata est hec Cosmographia que Scoti dicitur[3] cum picturis ex uetustissimo codice quem habui ex Spirensi bibliotheca anno domini m.cccc.xxxvi. mense Januario. Dum ego Petrus Donatus dei paciencia episcopus Paduanus uice Sanctissimi Domini Eugenii pape iiii Generali Basiliensi Concilio presiderem.

*ex Spirensi bibliotheca ... dum ... Basiliensi Concilio presiderem*: does this imply that the archetype was sent from Speyer to Basel for copying?

3. M is a copy of 1542 made directly from Σ at Speyer. It was sent in 1549 to Heidelberg, to the Pfalzgraf Ott Heinrich, who had written in 1548 to ask the Speyer authorities to lend him Σ itself for transcription. Their refusal was minuted on 26 January 1548:

> Herzog Ott Heinrich pfaltzgraff schreibt vnd begert den

Itinerarium Anthoninj, sonderlich das alt Exemplar Ir Gn. zu leyhen, wol dasselbig abschreiben zu lassen ... Darauf Ir gn. geantwurt: das alt exemplar sey etwas blod vnd steen m. Hn. im werck das zu renouiren.

Ott Heinrich agreed to accept instead a copy to be made at Speyer, specifying that the pictures (in which he seems to have been primarily interested) should reproduce as accurately as possible those in Σ: and the dean of Speyer directed one Dominic to prepare the new manuscript. Nothing, however, had been done by December 1549, although the authorities held out the prospect of producing the copy when the spring came round. In the meanwhile, the dean attempted to pacify the importunate Pfalzgraf by the loan of an existing transcript of Σ *uf die newe hand gemalet*, namely M; and, the new copy still not being ready by July 1550 - a Speyer minute of 20 March discloses that no competent miniaturist had yet been found - the loan of M was converted into an outright gift, as M's prescript records:

HIC LIBER, CVI TITVLVS, ITINERARIVM ANTONINI, ad uerum atque archetypum exemplar descriptus, Illustrissimo Principi ac domino, domino OTHONI HENRICO, Comiti Palatino Rheni, ... a Venerabilibus ac honestis Cathedralis Ecclesię Spirensis Decano atque Canonico Dono missus est.

But Ott Heinrich refused to be satisfied, because (and here again I quote M's prescript)

eiusdem libri picturę ... praesentis aetatis habitum, ac nouitatis formam quandam pre se ferebant, veterisque atque archetypi exemplaris Schematibus ac lineamentis non ita exactę atque per omnia respondebant,

and it was arranged that a Master Hans should paint a separate set of pictures from tracings on oiled paper, if these could be made without staining Σ. The new reproductions, cued to the text with numbers, were placed at the back of M when it was rebound for Ott Heinrich in 1551; and with this rather unsatisfactory substitute for the *uerum atque archetypum exemplar* the Pfalzgraf had to make shift until the following year, 1552, when, as we shall see, he contrived to acquire Σ itself.[4]

**4.** V was completed in 1484, as its subscription proclaims. A clue to its origin is found in another Vienna manuscript, *uindobonensis* 3102, which exhibits the following colophon in the hand of Bernhard Cles:

> librum hunc satis incorrectum ... est iussu nostro transcriptum ex antiquo exemplari in bibliotheca capitulari spirensi ... anno 1529.

The *antiquum exemplar* was not Σ but V, with which *uindobonensis* 3102 agrees essentially *verbatim et literatim*, even to the extent of copying its subscription.[5] V therefore was in the Speyer library in 1529, and there is a powerful presumption that, like M, it was a direct apograph of Σ made for some unspecified purpose - perhaps to save wear and tear on the frail original. The supposition is confirmed by those of V's errors which show that it is independent of the earlier copies C and P -

> *d.r.b.* pr.12 *turbae* omitted in C, present in V
> *d.r.b.* ii.1 *sub* P *sed* V

- but that it transmits corruptions found in Σ (the lacunose *laus aut prosequatur* at *d.r.b.* pr.3) and was copied - it is necessary here to anticipate a later conclusion - from an original in insular script (*d.r.b.* vii.6 *intoxta* V, for *intorta*).

**5.** I turn now to P, of which Thompson says 'It is not altogether certain that P is a direct copy of the Spirensis, but, if not, its intermediary is now lost.' P undoubtedly transmits, under a layer of its own superficial corruptions, a text agreeing with that of Σ as reconstructed from the united testimony of C M and V. Thus, at *d.r.b.*pr.10 it inherits *alieni*, a *lectio difficilior*, and in *d.r.b.* xiv.3 the desperately corrupt text reads identically in all four manuscripts. That P, which was in existence by about 1443, is not a copy of the contemporaneous transcript C is evident from the presence in P of *turbae*, omitted by C in *d.r.b.*pr.12, and from P's correct *non ferenda* in *d.r.b.* iii.1, where C gives *conseruanda.*[6] That P is not a copy of an archetype independent of Σ is suggested by its taking over corruptions which may with reason be supposed to have originated in Σ itself - the dittography *et minori ut minori* at *d.r.b.* v.7, the loss of a termination in *d.r.b.* xv.4, the spelling ASCOGEFRI in the picture-caption for *d.r.b.* xvi. It is therefore conceded that P derives from Σ either directly or through an intermediary which was not C; and the case can, I think, be decided on the basis of three readings. At *d.r.b.* xii.3 *erigunt* P's scribe at first mistook his exemplar's *r* for an *x*, subsequently correcting it over the line; at *notitia* or.xii.27 P's original *cyrecei* is similarly corrected to *cynecei* (read, of course, *gynaecei*); and, most revealingly of all, P twice (*notitia* or.xxiii.2 and 3) writes *libra* for *libya* without correction. The confusions are eminently characteristic of transcription from an insular hand, and it is clear - the demonstration follows in §16 - that at least the *notitia dignitatum* and its satellite texts, the *de rebus bellicis* amongst them, appeared in the *spirensis* in an insular script. P therefore was transcribing from an insular book carrying a text identical with that of Σ: and this must surely mean that P was transcribing from Σ itself.[7]

6. I proceed to the question of the exemplar used by Ghelen in preparing the first printed edition of the text. Here, as nowhere else among the problems posed by study of the Speyer *notitia* corpus, *grammatici certant et adhuc sub iudice lis est*, a posture of affairs for which the learned drudge of Basel is himself partly to blame. It seems to have become an article of faith, despite acknowledged difficulties in Ghelen's title-page and preface, that his text depends

somehow upon the same *spirensis* which is the archetype of our surviving manuscript copies; the sole ground for this supposition being that this is the only ancient manuscript containing all the works which he prints of which we have any cognizance. This casual (but, as it transpires, correct) assumption should not have been allowed to go without proof, and a substantial part of the following discussion is devoted to establishing that Ghelen did indeed draw on the Σ tradition of the collection. The argument will then be driven to a conclusion which seems, on internal and external grounds, both logical and inevitable; which was contemplated (without adequate demonstration) by Seeck; but which Thompson (who explicitly shrank from the fundamental task, that of inquiring for resources of fact beyond the work with which we are immediately concerned) denied to be possible.[8]

7. Our first step, before we enter on the specific problem of Ghelen's 1552 *notitia* volume, must be to marshal a number of facts which will help us towards an understanding of the man and his methods. Between 1532 and 1554 there poured from the printing-house of Froben at Basel a flood of texts edited by *Sigismundus Gelenius*, Siegmond Ghelen [1497-1554], *uir utriusque litteraturae non uulgariter callens*, as the printer's puff in his Ammianus calls him, *emunctaeque naris*. They make a formidable catalogue. Aurelius Victor with Eutropius and Paulus Diaconus appeared in 1532, and Callimachus' hymns *cum scholiis nunc primum editis* in the same year, followed in 1533 by Ammianus Marcellinus (in a volume which also included the *historia augusta*, Suetonius, and a Latin Dio) and a *geographici graeci* in which Ghelen collaborated with Arnold van Arlen. In 1535 he printed his *castigationes* of the Elder Pliny, and produced an edition of Livy jointly with Bilde von Rheinau. Two years later Ghelen published a faintly eccentric philological work, a *lexicon symphonum* ('*quo quatuor linguarum, graecae scilicet latinae germanicae et sclauinicae*[9] *concordia consonantiaque indicatur*') and then disappeared for a while, perhaps so that he could build up

a stock of fresh material. A second edition of the *lexicon* was called for in 1544, and his Arnobius *aduersus gentes* was printed in 1546. He now began to concentrate on Greek authors: both Aristophanes and John Chrysostom appeared in 1547, the latter perhaps in collaboration with Simon Gryné, Greek professor at Basel, whom we shall meet again in another connexion and who, interestingly enough, had edited Aristophanes for Brubach of Frankfurt three years earlier. Ghelen's Latin Josephus came out in 1548 - van Arlen had edited the Greek text in 1544 - and his *anthologia palatina* (reprinting the notes of Brodeau) in 1549, the year also of his Dionysius of Halicarnassus.[10] He returned to Latin in 1550 with Tertullian; and a gap of two years intervenes between this and the *notitia dignitatum* of 1552, after which there ensued another pause until 1554, the last year of Ghelen's life, an *annus mirabilis* which saw the publication of his Appian, his Philo, and his Elder Pliny - this last a *Lebenswerk* to which the *castigationes* of 1535 had been a prelude. He had also found time to work on Ambrose; the edition appeared posthumously in 1555.

8. This is not an unimpressive achievement; but it has a rival in the output of another of Froben's client editors, Bilde von Rheinau, *Bildius*, later - and more grandly - *Beatus Rhenanus* [1485-1547], Ghelen's inseparable colleague, who, for reasons which will appear, has a certain claim on our attention. He began production gently in 1509 with editions of Faustus Andrelinus and Michael Tarchaniota: deserting Basel for Strasbourg, he published Pliny's letters there in 1514, but returned to Froben's *ergastulum* in time to issue in 1515 a somewhat irresponsible collection which comprised Seneca's *apocolocyntosis*, Synesius *de laudibus caluitii* and Erasmus' *Moriae encomium*. His *Germania* of Tacitus appeared in 1519; Velleius Paterculus *ab interitu uindicatus per Beatum Rhenanum*, and the *panegyrici latini*, followed in 1520. In 1521 he edited a compilation of *traditiones de arte rhetorica*, and in the same year a Tertullian: another collection, *auctores historiae ecclesiasticae*, came out in

1523, and a set of corrections in the Elder Pliny in 1526. Finally, in a burst of chalcenteric activity between 1531 and 1536, Bilde produced editions of Prokopios (1531), Livy (1535; in collaboration with Ghelen), Tacitus (1536) and Origen (1536), as well as an augmented version of his *auctores historiae ecclesiasticae* (1535). His last years were spent in preparing revised editions of his earlier work, and of these the most important is the *editio tertia* of his Tertullian (1539), for which he was able to use a new and valuable manuscript source.

9. Bilde, despite his enormous productivity, seems to have been something of a serious scholar; Ghelen, because of his, partakes more of the character of printer's hack, equally prepared to supply, as occasion required and at the shortest notice, copy for a Latin historian, a Greek dramatist, or a glossary of homonyms.[11] But the relationship between the two men was evidently very close: they shared the same interests, they worked for the same publisher; they had at sundry times become involved in actual collaboration, whether overt (as in the case of the 1535 Livy) or implicit (Bilde's emendations in the Elder Pliny of 1536 being taken up by Ghelen in his castigations of the same in 1535 and his edition of 1554); and Ghelen brought out a new edition of Bilde's admired Tertullian three years after his friend's death.[12]

Another fact connects them. In the preparation of many of their editions they were able to employ newly-discovered manuscripts which permitted them occasionally to present items for the first time, more often to improve dramatically the current text of known authors. The provenance of these manuscripts is usually revealed in the prefatory matter to the various publications. Thus, Bilde's Velleius of 1520, the *editio princeps*, was printed from a book found at Murbach: this library also supplied an Elder Pliny whose superior readings were reported by Bilde in his critical study of 1526 and which may have been directly available to Ghelen for use in his subsequent work on the text. Ghelen's Ammianus of 1533 was based on a manuscript from Hersfeld, which almost completed and very much purified the mutilated and depraved vulgate.[13] The collaborative Livy of 1535 contained, for the first time, the fifth decad, taken from a manuscript found at Lorsch in 1527 by Simon Gryné; for the third decad, moreover, the editors were able to use a Speyer text, and, for the fourth, another book from Speyer and one from Worms. The new (1530) edition of Bilde's Plinian emendations states that *tres noui codices* (they cannot now be identified) have been applied to the revision of the work; the 1535 print of his *auctores historiae ecclesiasticae* included a *princeps* (that of Nicephorus) from an unknown source; and the third edition of his Tertullian (1539) incorporated new readings from a manuscript discovered at Gorzé.

10. These excellent and (presumably or demonstrably) old manuscripts, found and used by Bilde and Ghelen between 1520 and 1539, all belonged, so far as we are told, to cathedral libraries in a tolerably well-defined area of the upper Rhineland, not excessively far from Basel. It is interesting that they should all have become available at this late date and so closely together. Bilde and Ghelen, or their agents, had evidently been exploring the undisturbed resources of the neighbouring monastic collections. Nor were they alone, for, as we saw, Gryné, another Basel scholar, had penetrated into Lorsch in 1527; and Erasmus, for his 1518 Froben *historia augusta*, used a Murbach manuscript which is now lost. In patronizing the work of these men the Frobens, father and son, appear to have been aiming at the publication of two particular varieties of classical text: on the one hand, privileged and vastly improved editions of existing works, and on the other, *principes*. The reputation of the publisher, no less than that of the editors, may have materially assisted the processes of discovery; for the 1533 Ammianus, *exemplar manu scriptum gratis et alacriter suppeditauit egregius princeps,*

*abbas hersfeldensis*: and more than one abbot may have provided a manuscript promptly and without fee when it was known why, and for whom, it was required.

But this is to speculate. What is known with distressing certainty is that all the above-mentioned manuscripts used by Bilde or Ghelen or both - two *spirenses*, a *laurishamensis*, a *hersfeldensis*, two *murbachienses*, a *uormatiensis* and a *gorzeiensis* - are now wholly lost, with the single exception of six leaves of the Hersfeld Ammianus. Their loss has serious consequences for the criticism of the texts concerned. How reliable are the editorial reports, and to what extent are the printed texts sophisticated by unsignalized conjecture? The problem is most acute in the case of Velleius, where Bilde's *murbachiensis* was an *unicus* and, that having disappeared, the print is *codicis instar*:[14] and the hideous truth is revealed in the case of Ammianus, where it is still possible to confront the printed text with portions of the manuscript from which it derives. The leading characteristics of Ghelen's editorial technique will appear from what follows, as we examine his *princeps* of the *notitia dignitatum* and its associated texts - a work which has been deliberately excluded from mention in the survey above.

**11.** The contents of the manuscript used by Ghelen in the preparation of his *notitia* volume of 1552 are given on his title-page[15] in this order: *notitia dignitatum*, *descriptio urbis Romae*, *descriptio urbis constantinopolitanae*, *de rebus bellicis*, and - *ne quid de antiquo exemplari omitteretur* - the *altercatio Hadriani et Epicteti*. The actual printed order, given and discussed below at §**21** *init.*, differs from this, and in neither case are we entitled to assume that the sequence of the items in the exemplar has been retained. Now these items, and these only, form the contents of two surviving manuscripts, *uaticanus latinus* 3715 and *leidensis uossianus* 44: but neither can have been Ghelen's copy-text, since both are demonstrably copies of items selected from C, the Oxford transcript of the *spirensis*, and we shall see in the next paragraph that Ghelen could not have used C or any manuscript carrying C's text.[16] The identity of contents between these two books and the printed text is interesting; we shall recur momentarily to it: but the nature of their text precludes any attempt to identify one or other of them as Ghelen's exemplar.

**12.** The argument from contents having failed us, we are thrown back on the evidence of readings. Let it now be demonstrated that the text of the items which Ghelen prints was derived from none of the surviving direct transcripts of the *spirensis*. Confining our attention merely to *de rebus bellicis* and a few pages of *notitia dignitatum*, we find conclusive evidence that the manuscript used was not C:-

*d.r.b.* pr.12 *turbae* omitted by C, present in Ghelen and MPV
pr.14 *pauci* C *perpauci* Ghelen and MPV
*notitia* oc.xxix.14 *ceteros* C *reliquos* Ghelen and MPV
or.vi.68-9 omitted by C, present in Ghelen and MPV

Nor was it M:-

*d.r.b.* iv.5 *auidi* M *cupidi* Ghelen and CPV
*notitia* or.xxxiii: entries 24-28 transposed to follow entry 35 in M, correctly placed in Ghelen and CPV

Nor P:-

*d.r.b.* iv.1 *existimabant* P *existimant* Ghelen and CMV
iv.2 *dirigit huiusmodi* P *huiusmodi dirigit* Ghelen and CMV

Nor V, since V uniformly omits the illustrations, which are present without exception in the printed edition: and if it were contended that Ghelen might have derived his pictures from another source but his text from V, the supposition would be demolished by such verbal evidence as:-

*d.r.b.*pr.10 *igitur* V *enim* Ghelen and CMP
i.7 *utilitatis* V *uilitatis* Ghelen and CMP.

In all the above cases Ghelen gives the true reading, and in none of them, I think, could critical sagacity or temerity have detected the fault, or, having detected it, so infallibly removed it.[17]

**13.** Ghelen, then, did not use C M P or V as copy-text, and *a fortiori* he did not use any of their progeny. But that his text bears in erroneous readings a conspicuous and detailed similarity to that of Σ as it is reconstructed from the direct copies there is equally no doubt:-

*d.r.b.*pr.13 *eques* Σ, Ghelen: read *equus*
vii inscr. *exemplum* Σ, Ghelen: read *expositio*
viii.5 *extremitate* Σ, Ghelen: read *extremitates*
*notitia* or.i.38-9 *duces per* Σ, Ghelen: read *duces xiii, per*
or.ii.1 *uirorum illustrium praefectorum* Σ, Ghelen: read *uiri illustris praefecti*
oc.xi.74-77 *branbaricariorum* Σ, Ghelen: read *barbaricariorum.*[18]

And where Ghelen's text does differ from that of Σ his minor divergences, where they are not mere spelling-corrections, can invariably be explained as more or less intelligent and often successful conjectures based on corruptions identical with those carried by Σ:-

*d.r.b.*pr.7 *prope* Σ *propter* Ghelen
xviii.7 *communite* Σ *communitam* Ghelen
*notitia* or.xxviii.19 *traian(a)e* Σ *traiana* Ghelen
or.xxviii.31 *touia* Σ *iouia* Ghelen

I would without hesitation explain his major variants also as the product of deliberate alteration of a *spirensis*-type text: a representative selection of them is discussed in §**17** below. In no case need we postulate that Ghelen may have been controlling a text of the Σ-type by an independent manuscript source.

**14.** Important evidence for the nature of the text used by Ghelen appears from the *descriptio urbis constantinopolitanae*, printed in the *notitia* volume and contained in Σ. The text of this little work is separately preserved in a manuscript written in about 940, probably at Fulda, namely *uindobonensis* 162 - I will call it U - which is demonstrably independent of the *spirensis*.[19] From c. ii.12 of the text onwards U is substantially in agreement with Σ in presenting the readings of their ultimate parent, whose corruptions (as at viii.17) and omissions (as at xi.2 and 8) are reproduced in both manuscripts. U, however, had evident difficulty in deciphering the handwriting of its exemplar, which was in an insular script, as the characteristic misreadings show: iii.4 *mane* for *mare*, iv.2 *tertio* for *tertia*, iv.10 *seminotundam* for *semirotundam*, iv.16 *magnus* for *magnas*, and notice the correction at viii.6-7, *p°articibus*. That the exemplar of U was not Σ (on whose script see §**16** below) is shown in this portion of the text by the *habet* given by Σ in vi.18, of which interpolation U is innocent, and by such divergences as *confinio* Σ, *conuiuio* U at xiii.5. But the most instructive passage of text is at the beginning, Seeck's cc. i-ii, which are quite differently preserved in the two branches of the tradition. U is alone in exhibiting (though not with outstanding accuracy) c. i: nothing of this, or of most of the opening of c. ii, appeared in Σ, which presented a mutilated or deliberately abbreviated version of the text, beginning *regio prima continet* (*continet* om. V) *domum Placidiae augustae*, where *regio prima continet* looks like an interpolation and *domum Placidiae augustae* falls at line 11 of c. ii in Seeck's edition. Σ, for its part, preserved one feature entirely absent from U, the picture at the head of the work. This, it is true, survives only in

C, but blanks are left for it in M and in V (which omitted all the illustrations on principle), and C was not therefore merely inventing an appropriate frontispiece where none had existed.

Comparison of Ghelen's text of the *descriptio* with the variant versions presented by Σ and U reveals that it follows the tradition of the *spirensis* throughout. It begins *regio prima, continet in se* (an editorial supplement)[20] *domum Placidiae augustae*, and thereafter exhibits the peculiarities of Σ (vi.15 present in Σ and Ghelen, omitted in U: vi.16 *ualentiana* Σ and Ghelen, *ualentiaca* U *recte*: vi.18 *portum habet prosforianum* Σ and Ghelen, *habet* om. U), sometimes modifying the text by conjecture (so the supplement mentioned above, and e.g. iii.1 *post* ΣU *per* Ghelen[21]). And Ghelen prints before his text a picture agreeing - subject to the limitations of Froben's engraver - with that copied from the *spirensis* by C alone.

**15**. It is thus very clear that for the works which he printed Ghelen had access to a manuscript carrying their text as is appeared in the *spirensis*. That he did not depend on any of the copies C M P V, or on any copy of these copies, was shown earlier, and some explanation other than that of direct dependence on C must be found to explain the presence in the print of the *descriptio* illustration preserved in that manuscript alone.

Certain facts may now be presented concerning the *spirensis* itself. This book, found first by Pietro Donato in the Speyer library in or before January 1436, and last copied there when Master Hans traced the illustrations in 1550, remained at Speyer (so far as is known) until 1552, when the Pfalzgraf Ott Heinrich, already the owner of M, swooped on the cathedral library and carried off the *spirensis* in triumph to Heidelberg. Adjacent entries for Σ and M appear in the Heidelberg catalogue of 1556.[22] Ott Heinrich died on 12 February 1559, and part of his collection was eventually removed to the library of his heir Wolfgang at Neuburg. The books so transferred were listed in October 1599, and among them were both Σ and M. The list[23] identifies and distinguishes between them by their physical appearance: Σ, the *itinerarius Antonini, das alt recht exemplar,* was *uf Pirgament geschrieben, mit alten illuminirten Figuren, gebunden in Bretter, mit weissem Leder überzogen*: while M, the *bemellter Itinerarius Antonini, uf Pirgament neu abgeschriben und illuminirt, durch Dechant und Capitl zu Speyr Pfaltzgraf Otthainrichen verehrt,* cut a more brilliant figure, *in Bretter gebunden, mit goldtfarben Leder überzogen, uf dem Schnitt und Leder vergult und mit messen Spangen beschlagen.* In 1660 M migrated to Düsseldorf, thence to Mannheim, and finally to Munich; Σ however was destroyed, whether accidentally or deliberately is not known. A single sheet, plausibly ascribed to it, survives as the former binding of a record-book of 1602/3 in the archives of Wallerstein, near Neuburg.[24]

The *spirensis* contained a remarkable collection of items of various dates, all of a geographical, topographical or more-or-less official character. They were, in this order, and with the titles suitably adapted to modern forms of reference,

Aethici cosmographia
itinerarium Antonini
Dicuil de mensura orbis terrae
notitia Galliarum
laterculus Polemii Siluii
de montibus portis et uiis urbis Romae
de rebus bellicis
altercatio Hadriani et Epicteti
descriptio urbis Romae
descriptio urbis constantinopolitanae
de gradibus cognationum
notitia dignitatum.

The list divides itself neatly into two parts: first, a set of topographical treatises, without illustrations and ranging widely in date;[25] and second, beginning with *de rebus bellicis*, a group of six items with official connotations,[26] all illustrated, and illustrated in a style consonant with that of the late fourth or early fifth century, a period to which the composition of some of the pieces can be securely ascribed. This second group clearly represents the contents of a single manuscript in which the separate works had been copied up together: the texts in the first group may not have been brought together before the compilation of the corpus which the complete *spirensis* represented.

When and how was this synthesis achieved? Seeck evolved a theory that the collection was assembled either by the scribe of the *spirensis* or by the scribe of its exemplar (which cannot have preceded it by any great distance in time, since the *spirensis* was, according to Seeck, no later in date than the eleventh century, and its exemplar would have had to contain Dicuil *de mensura*, which is firmly dated to 825). On either supposition, Seeck's view compels us to envisage the *spirensis* as an organic whole, a book written throughout in a caroline minuscule which employed the open *a*, its text already plagued with insular corruptions which had originated at some point in a long and complicated history of copying and which are faithfully reflected in C M P and V.[27]

**16.** But I submit that this was not the case: for it seems to me quite clear that in the group of fourth- or fifth-century texts isolated above from the contents of the *spirensis* (the *notitia*-group, as I shall call them) the direct copies of the Speyer book, C M P and V, were themselves transcribing from an insular hand. This was briefly canvassed for P in §**5** above, and is discussed in detail for all four manuscripts at the appropriate points in my critical apparatus. The crucial readings in *de rebus bellicis* are these:-

> ii.6 *tyrannos* corrected from *trrannos* in V
> vii.6 *intoxta* V *in auram* M *intorta* CP
> xii.3 *exigunt* (with *r* written above the *x*, *x* marked with a dot beneath) P
> xix.3 *ascogefrri* C.

These errors point unmistakably to an insular exemplar; and observe that in these instances we are not dealing with four scribes each blindly copying the same existing corruption, a word already mangled in Σ: they are independently attempting to decipher the crabbed, unfamiliar and perhaps faded script of an insular book. Examples of the same confusions of letters can be found in one or more of C M P and V throughout the *notitia*-group; and we are, I think, obliged to conclude that the text of the works in this group at least was written in the *spirensis* in an insular hand. But Σ may not have been an insular manuscript throughout, for, if the leaf of the *itinerarium Antonini* found at Wallerstein is correctly recognized as a fragment of the *spirensis*, then that work at least, and perhaps the entire first portion of the manuscript, was written in a late ninth- or early tenth-century caroline, allegedly of Middle Rhenish type. And even without the evidence of the Wallerstein sheet we might precariously surmise from the corruptions of C M P and V that in the works outside the *notitia*-group they were not transcribing from an insular original: none of the typical confusions of *n* and *r*, *x* and *r*, *y* and *r* are recorded,[28] despite quite unusually prolific opportunities for their occurrence in unfamiliar geographical names. We must therefore suppose that the *spirensis* was not an organic whole; that it consisted of at least two groups of manuscript leaves, the *itinerarium* and the geographical works (if they were copied up with the *itinerarium*) in caroline minuscule, and the *notitia*-group in insular minuscule, bound together to form a single book.[29]

**17.** The connexion of Ghelen with this book may now be investigated. We have seen (§13) that collation of his text reveals that it is basically in accord with that of the *spirensis*, and it was suggested that *all* variants in the print, of whatever magnitude, are the result of editorial conjecture. This will require extensive demonstration; let *de rebus bellicis* and the first few chapters of the *notitia dignitatum* serve as test cases.

We will begin with some of Ghelen's successes. At *d.r.b.* pr.17 he could correct *prope* (so Σ) to *propter* without difficulty, and Σ's *communite* was easily rectified at xviii.7. *portatilis*, which Ghelen prints for the corrupt *portanti* at xviii.8, is self-evidently conjectural and, I judge, right: but the *libido coniectandi* overcame him at v.6, where his *limites Arabum* for *limites, arabunt* is an emendatory exploit worthy of the mature Bentley; and his *sine* for *sane* in xix.2 betrays a woeful (but probably not wilful) misunderstanding of the drift of the author's meaning. Σ's corruptions in pr.3 *nec te,*[31] pr.13 *eques,* viii.3 *extremitate,* and the defects in xiv.3, all eluded him - perhaps fortunately, since their appearance in his text shows that his exemplar reflected the most minute peculiarities of the Σ tradition. He did his best with the wretched *alio* in v.4, accenting it as an adverb (with no perceptible improvement in the sense); and he had the tact not to meddle with *uenientum* in ix.1.[32] He rightly deleted the dittographic *ut minori* in v.7, and I acquiesce in his removal of *et* from the headings of c. x. In neither case does Ghelen give any indication that he has eliminated matter from the text, but we are not therefore to assume that he was following a manuscript which carried a version of the text modified from that of the *spirensis,* for the silent deletion of difficulties is an outstanding characteristic of the Gelenian critical method. Just so he excised *aut* from the received *laus aut prosequatur* in pr.3, where we nowadays perceive the loss of a word before the verb. There are convincing parallels in his treatment of the texts of Ammianus[33] and the *notitia,* where a spectacular deletion will be commented upon shortly.

Turning, then, to the opening chapters of the *notitia,* we find ample evidence here also that Pasquali's *congetturatore abilissimo e praticissimo ... ma anche altrettanto audace* is at work on a text of purely Σ-type. *praesentalis* for *praesentat* was obviously necessary at or.v.26 (the mistake had probably resulted from the misapprehension of the abbreviation *praesentat*-), and the supplement IX at or. vii.38 required of its restorer only the ability to count. But he was not so fortunate with *consularem* for *consularitatem* at or.i.78, or with *Aegyptum* for *aegyptiacam* at or.i.80; and *graecorum* for *graecarum* at or.i.24 is the result of a faulty articulation of the context. In or.i.40 his *Libyae* for the important corruption *librarum* misses the obvious truth with almost deliberate perversity;[34] and *magistri* for *magister* at or.v.75 is wholly unprovoked. From his reading of or.xxiii.2-7, 9-14 he inferred a lacuna after or.i.29 which he filled with the apparently missing name *Augustamnica* - an alteration which in turn necessitated the change of the numeral in or.i.24 from V to VI. As in *de rebus bellicis*, so in the *notitia* some genuine difficulties escaped him, the missing entry after or.vii.34, for example; and patent incoherencies were concealed by silent deletion, as in the illustration to or.vi, where entry 24, a duplicate of 22 as regards the caption but without a coloured device on the shield, is suppressed. There are certain curiosities. In one entry (or.i.5) Ghelen adds a marginal note, cued in the text with a dagger: text †*in praesenti ii*, margin *alias, praesentiales siue praesentales*. Either he thought emendation unnecessary here, or, more probably, he was simply drawing attention to the unusual terminology: he is certainly not reporting a variant from another manuscript. And this is not all: immediately following the numeral *ii* what appears to be a jotting made by the editor in the printer's copy has actually crept into the text, which reads, in full,

MAGISTRI equitum & peditum †in praesenti II. Alij III.

But this is by no means the full extent of Ghelen's interference with the text; when he is confronted with an exigency of typography rather than of criticism, his expedients can be shockingly arbitrary. Thus, *notitia* or.ii.59 is followed in the printed book by or.iii.4 sqq., the series of entries ii.60-71 being entirely cut out and the illustrations at the head of c.iii temporarily deferred: moreover, iii.33 is replaced by ii.72 with its text ruthlessly altered and interpolated (*uterque praefectus ... ipsi praefectos emittunt*) to suit the faulty plural *uirorum illustrium praefectorum* of iii.1, 4 and 20, which had arisen in or before Σ from the wrong expansion of an abbreviation. By these brutal means the letterpress of cc. ii-iii is neatly concluded at the foot of sig. b1r, leaving b1v free to carry the full-page plate of the insignia of the *praefectus praetorio per Illyricum* with its caption. Sig. b2r contains only the small plate of the other illustration to c. iii (*Macedonia* and *Dacia*); b2v and b3r are necessarily occupied by the illustrations and captions to c. v;[35] and the printed text resumes with the lists of c. v (Seeck's 26 sqq.) on b3v. These violent alterations have been carried out with a certain low scholarly cunning, and they are to be imputed not to Froben's caster-up or compositor but to the learned editor, who must have been present in the printing-shop when the copy was being marked for composition.[36]

**18.** I have dwelt longer on Ghelen's differences from the tradition of the *spirensis* than on his resemblances to it; but the essential similarities are sufficiently obvious, and what required demonstration was that whenever he diverges from the Σ-text we are nowhere entitled to conclude that he was using a manuscript independent of Σ, or even that he had more than one manuscript at his disposal. What, then, was the manuscript which he used? That it must have been **the spirensis itself** I would argue from strong external and admittedly weaker internal evidence. We have seen (§§ **9–10**) that for some time before the appearance of the *notitia* volume in 1552 Ghelen and Bilde von Rheinau had been exploring the monastic libraries of the Upper Rhine, amongst them Speyer; and the *spirensis* of the *notitia* had been located and publicized in 1525 by – Bilde von Rheinau himself.[37] It is inconceivable that his sinister crony Ghelen did not know of the manuscript almost from the moment of its discovery. What is curious is that neither scholar printed it immediately. Permission to do so might at first have been refused because of the fragile condition of the manuscript alleged in 1548; but both Bilde and Ghelen seem to have had their methods for the eventual extraction of old copies from recalcitrant authorities. In any case, neither was unoccupied at the time: the new *murbachiensis* of the Elder Pliny claimed Bilde's attention in 1536, and Ghelen, the junior partner in their enterprises, had arrived in Basel from Venice only in 1524; he was perhaps not to be trusted with a *princeps* until he had shown his paces for Froben in a less responsible task.

Both men must have been interested in the discovery by Simon Gryné, in 1527, of the *laurishamensis* of Livy's fifth decad; and when subsequent researches at Speyer bore splendid fruit in the shape of texts of decads four and five - the *uormatiensis* of decad four being also to hand - the production of a new Livy must have taken precedence over much else. Moreover, by 1533 Ghelen had secured the *hersfeldensis* of Ammianus and was working, for Froben, on an edition to rival Accorsi's and replace that of Erasmus.[38] These tasks completed, he could turn to the *notitia*: his manuscript was certainly available to him in 1536, when, as we shall see in the Excursus below, Froben's artist cut one of the woodblock illustrations for the book. Another picture, however, as we shall also see, was not engraved until 1550; the edition appeared in 1552; and (since the interval between 1550 and 1552 poses no problem, as allowance has to be made for the cutting of perhaps more blocks than one and for the setting of the letterpress) we are left to find a reason for the gap between the illustration dates. Infrequent access to the fragile and jealously-guarded Speyer exemplar is the obvious explanation of this, as it might also in part explain the eleven years' delay between the location of the manuscript and the cutting of one picture-block in 1536; though appeals to the bad condition of the *spirensis* inevitably depend on the veracity of the authorities in informing Ott Heinrich that their book was in a poor state, necessitating actual repair in 1548. An insular manuscript might well have been falling apart by then, but the possibility remains that Speyer's reply to the Pfalzgraf was an excuse concocted to try to keep the book out of the hands of an acquisitive bibliophile (and perhaps of destructive editors also). The probability of a link between the condition of the *spirensis* and the time taken to produce the printed text will be noticed in the Excursus at the end of this section: here I will only insist that Ghelen's strong interest in Speyer, his proven use of two other *spirenses*, and his presumed knowledge since 1525 of the existence of the Speyer *notitia*, are valid enough external reasons for believing that this manuscript was the direct source of his text.[39]

**19.** That Ghelen made direct use of the *spirensis* in preparing his edition of the *notitia*-group texts may also, I believe, be sustained from the internal evidence of two unusually interesting readings in *de rebus bellicis*. At the end of c. vii.2 Σ's writing of the word *intorta* was (or had become) sufficiently indistinct to confuse two of the scribes copying from it. V misread the insular *r* as an *x*: M, bewildered by the shapes of round *t* and open *a*, and unhappy about the end of the word, produced the wildly wrong *in auram*.[40] C and P wrote the word correctly: is this a tribute to their superior discernment, or had damage occurred to *intorta* between the dates of C and P and that of V? Now Ghelen's text contains no hint of a word resembling *intorta*, and for *iaculatur* he reads *eiaculatur*. What does this mean? Evidently that he could not fully decipher the reading in his exemplar of the word between *radiis* and *iaculatur*.[41] Neither C nor P was his exemplar, as we have seen (and as this variant further shows); and he could surely have corrected *intoxta* if V had been his copy-text (which, as we have also seen, it was not). Either, then, Ghelen was using M, with its ghastly error *in auram*: or he was using a (lost) copy or twin of the *spirensis* which wrote the word indistinctly or wrongly: or he was using the *spirensis* itself. But he did not use M: lost transcripts are not to be postulated unnecessarily: and given the possibility that Ghelen could have had access to Σ between 1525 and 1552, no alternative need be supposed. Ghelen, I contend, saw the doubtful reading in the *spirensis*, and his solution of the problem is entirely typical: the unclear word is suppressed, and *iaculatur* is wilfully extended, perhaps to compensate for the preceding excision.

The second important reading is at xxi.1, where all four apographs of Σ present us with *componatis*. Gundermann is right in restoring *comparatis*; the error arose from an easy misreading of insulars. But it is, I think, unlikely that all four manuscripts would misread the same two letters in precisely the same way, although they are elsewhere guilty of committing the same fault independently in a single letter. *componatis* must therefore, on this view, have been the reading of Σ, an error which tends to show that Σ was itself transcribing from an insular hand.[43] But P, having duly transcribed *componatis*, felt misgivings about what he had written, and placed expuncts under the letters *na*, without, however, suggesting alternative interpretations (he cannot have intended a deletion). This is exceptionally instructive. P had no doubts concerning the *o* before the *n* (which incidentally reinforces the contention that *componatis* was in Σ); his dissatisfaction was with *n* and *a*, which, in transcription from insulars, is understandable; but he evidently came to the conclusion that that was what the letters in his exemplar really were. Now it is precisely these two letters, *na*, which are altered in Ghelen's reading here, *compositis*. Had Ghelen been able to read *comparatis* securely, even he could have had no reason to change it; but two crucial letters were in doubt in his copy-text, the same two letters which had puzzled one scribe copying from Σ; and Ghelen, concerned (unlike C M and V) to produce Latin, was forced to improvise on the basis of what could be read without difficulty. We are, I think, driven to conclude that Ghelen had before him the very manuscript which deluded C M and V and troubled P, namely Σ itself. It would be excessively rash to postulate a lost congener of C M P V intervening between Ghelen and the *spirensis* at this point, since this hypothetical manuscript, if it were in doubt as to the reading - and the probability is that it would not have been -, must either have given Σ's letter-forms in facsimile or have already interpolated *compositis* (by conjecture, since P's failure to offer alternatives for the letters *na* shows that it was unsafe to copy them as anything else). And a *gemellus* of Σ would have had to present exactly Σ's unclarity. It is altogether more economical to suppose that no such intermediary or twin existed.

It is appropriate at this point to return briefly to a notorious reading in *de rebus bellicis* which has already been mentioned, namely *uenientum* at the end of ix.1, shared by Ghelen with P and V against the *uenientium* of C and M. If my contentions as to Ghelen's copy-text are correct, then of course he took his reading directly from Σ, and P and V must have preserved the truth by accurate transcription (which admittedly is the last thing we expect from V) rather than stumbled upon it by a fortunate error.

**20**. It is a remarkable fact that, of the nine great Latin manuscripts used by Bilde or Ghelen or both between 1520 and 1552, seven have completely disappeared, and the other two are represented only by fragments. Bilde is known to have sent the *murbachiensis* of Velleius to the printer as copy; it may never have come back.[44] The *hersfeldensis* of Ammianus, which also seems actually to have been lent to the editor ('*exemplar manu scriptum...suppeditauit...abbas hersfeldensis*') may similarly have gone to the press: if so, it was at some time returned, since the six fragments of it now at Marburg were found serving as dust-jackets to account-books at Schloss Friedewald, near Hersfeld. The Speyer corpus, of which we apparently possess a single leaf, was found in its home library in 1552 by Ott Heinrich. Questions arise, to which the reader may supply his own answers. Does the partial or total loss of these manuscripts imply that they were *all* at some time used in or near the printing-shop, suffering an ill-treatment which resulted in, if not their immediate destruction, then at least some damage which hastened their eventual collapse? Who was responsible for this? Was the Speyer corpus, or a part of it, ever lent, even briefly, to Basel after 1525, as it may have been in 1436, when C was copied? In fact, are its known fragility and ultimate ruin further evidence that Σ was used by Ghelen?

**21**. Two problems remain. The first concerns the contents of the 1552 edition, which, it will be remembered, consist of *notitia dignitatum, descriptio urbis Romae, descriptio urbis constantinopolitanae, de rebus bellicis,* and - *ne quid de antiquo exemplari* *omitteretur* - *disputatio Hadriani et Epicteti*. Neither this, the title-page order, nor the actual printed order - *descriptio Romae, descriptio constantinopolitanae, notitia, de rebus bellicis, disputatio* - reproduces the sequence of these items in Σ; and, despite the profession that nothing has been omitted from the antique exemplar, the little memorandum *de gradibus cognationum*, which was placed in Σ between the *descriptio constantinopolitanae* and the *notitia*, is not printed, nor indeed are any of the works outside the *notitia*-group. These facts admit of the following explanations:

(a) *The order of the contents*. This need cause us no disquiet. A sixteenth-century editor would not have felt himself bound to preserve the arrangement of his copy-text, and bibliographical considerations show that the contents have been so ordered that the *notitia* falls in the very centre of the printed book, symmetrically surrounded by the remaining matter. The modern treatises by Bilde and Alciati, which follow the preface and are strictly speaking extensions of it, completing the preliminary quire, together with the *descriptiones urbium* (Rome preceding Constantinople, as in Σ), occupy 21 pages of type: then comes the *notitia*: and *de rebus bellicis* and the *disputatio* (these two texts also in their Σ order) complete the print in another 21 type-pages. As for the title-page order, Ghelen or his publisher was clearly intent on securing the greatest possible typographical prominence for what was considered the most important item in the collection: thereafter the titles are in the printed order of the works.

(b) *ne quid de antiquo exemplari omitteretur*. If the *spirensis* was (as we have argued) a composite manuscript, it *may* have been obvious to Ghelen that the insular leaves containing the *notitia*-group texts were more archaic in appearance than the caroline leaves of the preceding portion of the book, and he *may* have wished to put only the more ancient section into type. This is improbable. It is much more likely that he was concerned to print only the spectacular illustrated items, as the written copies *uaticanus* 3715 and *leidensis uossianus* 44 were content to transcribe these alone from C; the script of this section could have helped to confirm that these illustrated texts constituted an *exemplar* separate from the rest of the manuscript and of undoubted antiquity. It is barely possible that Ghelen was permitted to extract these leaves from Σ and use them for a while in Basel: this would add verisimilitude to his (or his printer's) proclamation of an *antiquum exemplar* whose contents were given in full.

But in fact one treatise from among the *notitia*-group had been left out, the *de gradibus cognationum,* which consists of a tabular illustration and a short block of descriptive text. Its omission may have to be explained as a conscious excision made by a printer pressed for space and unable to include even so short a work without upsetting the planned make-up of his book. Something in the preliminaries, one feels, might with advantage have been spared to make room for *de gradibus,* which could have been printed on two pages in the chosen format and in Froben's standard roman type; but we must accept that Ghelen or Froben had legitimate reason for prefixing to the *notitia* Bilde's fragmentary but patriotic essay on Illyricum and Alciati's somewhat etiolated observations on Roman public offices. And given this, it can be seen by correlation of the collation-formula and the list of contents[45] that no two pages of the book are vacant, and that insertion of an extra sheet (say in quire p, to bring it up to a ternio) would have spoiled the distribution of the text over the quires and into 21 type-pages on each side of the *notitia.* Addition of a sheet would have further entailed either finding another one-leaf treatise to print or leaving an unsightly blank. Deliberate omission at the printing stage seems therefore the most probable reason for Ghelen's not giving *de gradibus.* A remote possibility is that the page carrying this text may have been missing from the fragile *spirensis* when Ghelen used the book: remote, in view of the undoubted presence in Σ of *de gradibus* when M was copied from it in 1542. A final bibliographical speculation is best relegated to a note.[46]

22. The last remaining question concerns a sentence in Ghelen's epistle dedicatory which appears to convey information about the provenance of the manuscript which he used in preparing his text. It reads *in extenso* as follows: Ghelen is addressing the surgeon André Vesale, *Vesalius*, on sig. *2v of the printed book:-

> caeterum haec NOTITIA, dum Romana res staret, adseruata penes Primicerium notariorum, et tandem afflicto ac penè exciso imperio, cum caeteris spoliis in barbarorum manus delata, nunc demum ex ultimis Britannis antiquariorum studiis repetita, innotescit auspicio felici tui nominis, aeque clari ac gratiosi apud omnes quibus cordi sunt bonae literae.

It is difficult to know how seriously we should take what is proposed here, especially since Ghelen regards his manuscript as the actual working copy of the *notitia* - and I suppose of its satellites also - used in the office of the western *primicerius notariorum* in the early fifth century. Manifestly it can not have been. But what was his reason for believing that it came *ex ultimis Britannis*? Perhaps he recognized the insular script of the *notitia*-group leaves as characteristic of manuscripts written by the remotest Britons: perhaps there existed at Speyer some evidence, or at least a tradition, that Britain was the origin of this portion of the manuscript. The existence of such a tradition may be reflected in the alleged connexion of the first work in the Speyer corpus, the *Aethici cosmographia*, with the Irish scholar Moelbrigte, *Mariangelus Scotus* (1028-?1082): *hec cosmographia que Scoti dicitur*, as C, alone among the apographs of Σ, describes it. Moelbrigte left Ireland for Cologne in 1056, was at Fulda between 1058 and 1069, and spent the rest of his life as a recluse at Mainz: it is suggested that he brought with him from Ireland a manuscript of Aethicus at least, and perhaps also texts of other components of the Speyer corpus. The supposition, if correct, would be of signal importance, because, if Moelbrigte carried over a text of the *notitia*-group with the Aethicus, that text should have formed the very set of leaves which - it is conjectured - was bound up into the *spirensis*.[48] But a tradition of a British origin for this section of the manuscript may have persisted without reference to Moelbrigte's reputed connexion with the Aethicus, and in either case the tradition, if it ever existed, may have been mistaken. There is, after all, no overwhelming necessity to connect Σ's *notitia*-group portion, on grounds of content or script, with Moelbrigte or Britain at all. It may never have left Germany: for, if these texts formed part of a western archive, the separate originals of its contents may have been held at the administrative centre at Trier, and both the insular *spirensis* and its postulated insular parent[49] could have been written in the paradoxically-named *continental insulars* at any one of the Anglo-saxon foundations in Germany, at Fulda, for instance, at Hersfeld, or at Echternach. But here the piling of speculation on speculation becomes profitless, *e se' venuto in parte | dov' io per me più oltre non discerno.*

NOTES

1. This is *parmensis palatinus* 1280, saec. xvi (I.G.Maier, 27 *Latomus* [1968].96 *sqq.*, quoting P.F.Girard's identification on p.103 note 1). Panciroli gave merely a summary of the contents of *de rebus bellicis*, but presumably he knew it exclusively from the Parma manuscript.

2. According to R.Neher, in his Tübingen dissertation of 1911 on the *anonymus*, *parisinus* 5825E is a copy of P, *uaticanus barberinus* 157 a copy of M, *uindobonensis* 3102 a copy of V [this is certain: §4 above with note 5], and all the rest copies of C.

3. The scribe is of course using the heading of the first work in the collection as a general title for the book. On the implications of the reference to *Scotus* see §22.

4. I will transcribe here, for their importance in textual and iconographic studies of Σ and M, the extracts from the Speyer minute-book printed by Preisendanz in 1 *Zeitschrift für Buchkunde* (1924).15-16:

A (26 January 1548). *Itinerarius. Hertzog Ott Heinrich pfaltzgraff schreibt vnd begert den Itinerarium Anthoninj, sonderlich das alt Exemplar Ir. Gn. zu leyhen, wol dasselbig abschreiben zu lassen. Haben m. H. darum geredt vnd in ansehung, das die alten herren es nie mher außzuleyhen beschlossen, für gantz beschwerlich geacht, es Ir. gn. zu leyhen, wollen ehe Ihr gn. ein exemplar verfertigen lassen. Darauf Ir gn. geantwurt: das alt exemplar sey etwas blod vnd steen m. Hn. im werck das zu renouiren, so das vol pracht, wolle sie Ir gn. (wo anderst Ir gn. als dan noch daruff bedacht, ein Exemplar zu haben) eins vertigen lassen.*

B (3 February 1548). *Hat Hertzog Ott Heinrich wider geschrieben, das Ir. gn. zu friden mit einer Copeij, mit beger das die figuren wie die alten sollen gemalet werden, welcher muhe halb Ir gn. sich woll werden zu halten wissen; darauf soll Tymenick vicarj es in der Dechanej schreiben.*

C (5 December 1549). *Hertzog Ott Heinrich pfaltzgraff, hat des Itinerarij halb m. Hn. Dhumbdechan geschrieben vnd angemanet. Ist Ir F. G. geschrieben, wie das geschrieben exemplar vf die allt handt noch nit gemalet, man wolle vf den fruling, so es wieder warm, vleiß ankeren, das es gemalet; wo es aber ir g. benugig sein wolle, wolle man Ir g. ein exemplar vf die newe hand gemalet zustellen. vf solchs hat er wider geschrieben vnd begert, das man Ime wolle das gemalet zu sehen an schaden zuschicken. Ist bewilligt.*

D (20 March 1550). *Des Hertzog Ott Heinrichen vertrostung geschehen, den Itinerarium vf die alt hande malen zu lassen, vnd aber khein maler zubekomen, ist dem fabricken maister bevolhen sich besser vmbzuthun vnd wo kein maler zubekomen, deß dem Hertzogen zu berichten.*

E (1/2 June 1550). *Nachdem vf m. Hn. schreiben, das der geschrieben Itinerarius vf die alt hande zu malen, nit mag bestelt werden, Hertzog Ott Heinrich zu friden, das man Ime den Ihringen, so Ir. gn. hieuor besichtigt vnd vf die newe hande gemalt ist, vngebunden zuschicke vnd darbey anzeigen wolle, waß der rest, ist bedacht wiewol derselbe vf 69 gld gecost, Ir gn. solchen verfertigten Itinerarium zu schencken vnd Ir gn. den namen zu antwurten lassen, weß Ir gn. für sich selb dem Stifft zu gutten gebe, muße man geschehen lassen vnd zu Danck annemen.*

F (continues E). *Hat Hertzog Ott Heinrich wiederumb begert, dieweil der Itinerarius, so m. Hn. Ir gn. verert, zu wit vf die new handt gemalet, das m. Hn. vnbeschwerdt sein wolten, seine maler, so Ir Gn. derhalben geschickt, vf geoldrenckt pappeir, die alte figuren durchzeichnen zu lassen, vnd dieweil m. Hn. nit wissen, ob es on schaden geschehen moge, so ist bedacht, Meister Honsens Molerns Rhat dar zu zuhoeren. Wo es on schaden gesein moge, so ist es bewillget.*

Thompson's account of M seems to me confused.

5. *uindobonensis* 3102 naturally presents some independent readings, all of which, however, are evident distortions of unclarities in V.

6. For P's early history see Thompson, pp.7-8.

7. Those who regard *alieni* in *d.r.b.* pr.7 as corrupt would claim this as very strong evidence of P's descent from Σ and not from a *gemellus* of that manuscript, since the perversion of terminations was a characteristic of the Σ text (see my critical note *ad loc.*). If correct, as I believe it to be, the reading is less decisive.

8. On page 11 of his edition, lines 6-7. The enormous mass of material on parts of the tradition of the *notitia,* laboriously put together by I.G.Maier in 27 *Latomus* (1968).96-141 and in 28 *Latomus* (1969).960-1035, insofar as it reflects upon the source of the *editio princeps,* is susceptible of interpretation in more than one way.

9. Bohemian; Ghelen was a native of Prague.

10. This sudden spate of Greek activity is remarkable. Perhaps Froben, whose publishing house was (as Ghelen expresses it in his *notitia* preface) *literatorum commodis iam dudum deditissima*, had signified a desire to have some new editions of Greek authors among his stock; and Simon Gryné's recent appearance on the scene may also be significant. Not that Ghelen lacked Greek; before going to Basle in 1524 he had studied at Venice under Mousouros, whose teaching was admired by - amongst others - Bilde von Rheinau (see J.E. Sandys, *A history of classical scholarship* 2.79 note 4, for Bilde's punning encomium).

11. The kindest possible verdict on Ghelen as critic is perhaps that of Pasquali in *Storia della tradizione e critica del testo*[2] p.82: *il Gelenio, se in fondo era uno studioso onesto, era anche un congetturatore abilissimo e praticissimo di lingua e stile, ma anche altrettanto audace.*

12. But Ghelen based his text on a different manuscript, the fifteenth-century *uolaterranus* 298(5404).

13. Without space or adequate pretext to work out these contentions in detail here, I will record my belief that Ghelen's Ammianus was prepared not, as has occasionally been suspected, through heavy dependence on Accorsi's edition (which carried readings and the text of books 27 *sqq.* from the *hersfeldensis* or one of its close relatives), but was edited firstly, in books 14 to 26, on the basis of the 1518 Froben print controlled by the *hersfeldensis,* and subsequently, in books 27 to 30.9.6 (where *reliqua in archetypo desiderabantur*), by working directly from the Hersfeld text. Ghelen,

I think, supplied his printer with composite material from which to set up the new edition: a copy of the 1518 Froben (Erasmus' re-edition of Castelli's 1517 *editio bononiensis*) emended by collation with the *hersfeldensis*, plus a modified transcript from that manuscript of books 27 *sqq.*, a very long stretch of consecutive text unknown to editors before Accorsi and himself; the whole, and perhaps especially the latter portion, suitably revised on established Gelenian principles. Froben's desire to have editions of important texts more saleable than those of his rivals is evident in all this. Castelli edits (for Benedetti of Bologna) what was known of Ammianus in 1517, and Froben forthwith commissions Erasmus to re-edit (in fact, to do little more than reprint under a greater name) Castelli's text: Accorsi edits (for Otmar of Augsburg) an augmented Ammianus in May 1533, and two months later Ghelen's new Froben appears. Two months is a cruelly short space of time in which to collate, edit and put into print a text of about 125,000 words in a poor tradition, and very conceivably Ghelen had been working on Ammianus for some years, with knowledge of the Hersfeld manuscript; but the publication of Accorsi's sensationally-improved edition must have been the final stimulus to the production of a replacement of the now obsolete 1518 Froben. Of course Ghelen may occasionally have appropriated, or independently conjectured, readings which appear in Accorsi, or, for that matter, elsewhere: but I would prefer to consider him in general as a treacherous witness to the Hersfeld manuscript than as a shameless borrower who mangled what he took.

14. See *infra*, note 44, for a brief description of Bilde's copy for his Velleius; and in detail, Woodman's edition (1977) of Velleius 2.94-131, pp.3 *sqq.* of the Introduction.

15. I give a type-facsimile of the title-page of the *editio princeps* in my critique of previous editions of *de rebus bellicis*, printed *infra* as an appendix to this section. Before the ancient works Ghelen prints his own epistle dedicatory and two modern essays, Bilde on Illyricum and Alciati on Roman public offices: *vide* the contents-list in note 45, and, for the effects of the inclusion of these treatises on the casting-up of the copy for the book, §21.

16. *uaticanus* 3715, which today lacks about half the text of *de rebus bellicis*, actually copies out C's colophon, *exemplata est hec cosmographia e.q.s.* (*supra*, §2).

17. Critical sagacity would above all have been defeated in the case of V, whose text is shockingly corrupt throughout. Seeck should not have said, on p. ix of his edition of the *notitia*, that *apographa eius* (sc. *codicis spirensis*) *quattuor extant, quae* omnia *ea cum cura facta sunt ut etiam in minimis rebus conspirent*: this is scarcely true of M and grossly false of V. We may further discount the possibility (Thompson, p.12) that Ghelen's exemplar may have been *uaticanus* 3715, which alone states - a statement which recurs on Ghelen's title-page - that *de rebus bellicis* is addressed to Theodosius and his sons. *uaticanus* 3715 is a copy of C, and Ghelen's text is in detail not one which any copy of C could present (*supra*,

§12). Furthermore, his qualified statement - *uetustus liber DE REBVS BELLICIS ad Theodosium Aug. & filios eius Arcadium atq; Honorium, ut uidetur, scriptus* - implies guesswork; *ut uidetur* should not have been written if there was manuscript evidence for the address. Ghelen knew his ancient history, and merely made an unlucky inference from the *imperatores, imperator* and *filii* of the text: a mistake suggested perhaps by the references in the *descriptio constantinopolitanae* to *thermas arcadianas, domum Pulcheriae augustae, thermas honorianas, cisternam theodosianam* and so on. But note that, since Ghelen knew only the mutilated form of the *descriptio* text (*supra,* §14), the cue for the identification cannot have come from the direct allusion to Theodosius in the preface to that work (line 14 of Seeck's edition), a section of the text transmitted uniquely by *uindobonensis* 162 outside the Σ tradition.

18. Why did Ghelen fail to correct this?

19. It is called V by Seeck, who in his apparatus to the *descriptio* changes the siglum of *uindobonensis* 3103 from V to $S^1$ - an unnecessary and confusing alteration. *uindobonensis* 162 may be consulted in Unterkircher's photographic facsimile, *umbrae codicum occidentalium* 2 (Amsterdam 1960).

20. *continet in se* is the regular formula of introduction (cc. ii-vii, ix-xv): c. viii alone has *habet autem in se.*

21. Ghelen was much given to this sort of tinkering; compare his equally gratuitous *prae* for *pro* in *de rebus bellicis* i.9.

22. The entries are printed by Thompson, p.9.

23. Text of the following descriptions of Σ and M from Thompson, p.9.

24. *Fürstlich Oettingen-Wallerstein'sche Bibliothek, Schloss Harburg, MS.*I,2,$2^0$,37. The fragment is confidently ascribed to Σ by competent authorities (*apud* Thompson p.10 and Maier, 27 *Latomus* (1968).98) and it is very convenient, though by no means essential, to believe that their attribution is correct; see *infra,* §16.

25. The Speyer library seems to have been well provided with topographical texts. C, having copied (at Speyer? §2) the contents of Σ, adds the *demensuratio prouinciarum* and Cyriac of Ancona *de septem mundi spectaculis,* but admits, at the head of the *demensuratio,* that this work *non erat in praecedenti codice, sed de antiquissimo libro desumpta.* Was the Cyriac from the same source? and did the *antiquissimus liber* duplicate some or all of the Σ items?

26. The *altercatio Hadriani et Epicteti* can be described as an official text only by the exercise of extreme sophistry, and its presence in the *notitia*-group remains mysterious. I can only suggest that if (as I am inclined to believe) the collection was assembled in the scriptorium of an insular foundation from separate originals, some of them perhaps having survived as a bundle from an official source, then the didactic, even pedantic quality of the *altercatio* should

have appealed to men trained in the insular tradition, and the fact that the text is a para-philosophical dialogue with an emperor as protagonist may have helped to justify its inclusion in a technical collection of imperial character.

27. Seeck in 9 *Hermes* (1875).229, and in his edition of the *notitia*, p.ix with p.xxvi.

28. A sweeping statement, but I believe it to be true in the cases of those texts and manuscripts for which printed collations are available.

29. I cannot claim that the concept of the *spirensis* as a composite manuscript is new; Schnabel (29 *Sitzungsberichte der Preussischen Akademie der Wissenschaften, phil.-hist. Abteilung* (1929).242 *sqq.*, at p.257) devised it before me. His conclusion has been impugned by Lehmann and Thompson: I reached it independently, without consultation of Schnabel's article, and have left my text as I at first wrote it. *ualeat quantum ualere potest.* Comparison of Schnabel's position with mine will show that I accept a more orthodox grouping of the contents of Σ than he proposes; and I would tend to set the *terminus* for the writing of the *notitia*-group leaves quite one hundred years before his ninth-century limit. We need suppose only one intermediary stage between Σ and the individual archetypes of the *notitia*-group works; see the note immediately following.

30. Seeck's speculations as to the archetype of Σ fall of course. I do not find it necessary to assume, with him, a long history of copying of the *notitia*-group. The text of C - on the whole the best surviving copy of Σ - is of distinctly high quality, and some of its errors are undoubtedly self-inflicted; and between it and the separate originals of the texts in the group there need have intervened only one manuscript, in which the collection was first made, and whose presence in the background may be felt in certain corruptions discussed in my critical notes to *d.r.b.*pr.5 and xxi.1.

31. That this was a corruption already existing in Σ is suggested in my critical note to xxi.1.

32. The necessity for the reduction of the termination in clausula is discussed in my critical note to ix.1.

33. See *supra*, note 13, for an opinion on the copy for Ghelen's Ammianus. For his practice of silently deleting desperately corrupt or defective text in Ammianus see, for example, his proceedings at 28.4.26 (lacunose matter in MV omitted by Ghelen), 28.5.1 *dum* MV omitted by Ghelen (read *dux* with de Valois), 30.3.5 *sacramenti et inti* MV, *et inti* omitted by Ghelen (perhaps read *et ritus* with Clark).

34. *librarum* may have been the reading of Σ: see note 43.

35. Seeck's postulated c. iv is lost. It was missing from the *spirensis*, for C M P and V know nothing of it. But observe that in view of Ghelen's butchery of the text in this area it would be unsafe to conclude *solely* from his omission of 'c. iv' that he was using a text of *spirensis*-type.

36. I have treated this extraordinary piece of Gelenian misrepresentation in some detail because it must be understood that the telescoping of the text here is *not* due to Ghelen's having followed a tradition other than that of Σ but arises purely from the accommodation of text to typographical circumstances. His methods were not wholly arbitrary; they can be observed and defined (*supra*, §17 with note 33, and, for another aspect of his technique, note 21); but they were pervasive, and his inveterate habits of trivial (and often unnecessary) emendation, the silent deletion of difficulties, and abbreviation, must be constantly borne in mind by those dealing with any text in which Ghelen is to be employed as a witness to a *deperditus*.

37. Thompson, p.8; P. Lehmann in *Sitzungsberichte der bayerischen Akademie der Wissenschaften, phil.-hist. Abteilung* 1934.iv.12 *sq*.

38. Further consideration of Ghelen's Ammianus *supra* in note 13.

39. The timetable of Ott Heinrich's requests for the loan of the *spirensis* (*supra*, note 4) shows that Σ was ostensibly at Speyer continuously from January 1548 to June 1550; and in 1552 the Pfalzgraf descended on the library and removed the manuscript to Heidelberg. Thompson (p.11), rehearsing an argument of Seeck's, avers that the presence of Σ at Speyer 'during the years 1542 [the date of M]-1552' precludes Ghelen's having used it as copy. It does no such thing; indeed, it would rather encourage than preclude it. It is even a possible explanation of the authorities' unwillingness to let Ott Heinrich have the book that they were reserving it for Ghelen's use. Σ had been discovered at Speyer in 1525 by Bilde von Rheinau; it was still there in 1542 and from 1548. Throughout this period we have no evidence that it ever left Speyer, and since it is on the whole improbable (despite the possibility explored in §20) that so important and fragile a book would have been lent to Ghelen, restricted access to it in the library seems altogether the best explanation of the time taken in producing the edition. Ghelen's glancing reference to the fortuitous (and, by implication, recent) discovery of the text must be viewed with appropriate scepticism: *equidem cupiebam*, he says to Vesale, *iam ante extare monumentum aliquod mutuae nostrae beneuolentiae: uerum tua humanitas hoc quod primum fors obtulit, aequi bonique consulet* (ed. pr. sig. *2v).

40. The confusion of open *a* with *u*, though certainly *characteristic* of transcription from insulars, is not *confined* to copies made from an exemplar in such a hand. The confusion of *a* and *o* is slightly more suggestive, that of *a* and *t* noticeably more so, and the combined evidence has perceptible weight.

41. Typographical convenience did not have to be taken into account here.

42. It must be distinctly understood that when in this and later contexts I speak of a 'lost copy of the *spirensis*' I mean a transcript of Σ intended, as C M P and V were intended, to serve as an independent copy of the text with no ulterior purpose in view. I do *not* refer to any transcript which Ghelen might himself have made, or have had made, to serve as printer's copy if the use of Σ itself for this purpose was denied him. Whether the compositor set up his text from the actual pages of the *spirensis*, as Pickel's printer set up Hrotsuith in 1501 (*infra*, note 44), or whether he used a prepared transcript, or whether he worked from a transcript *plus* the original, as Bilde's printer did with Velleius (*infra*, note cited), cannot be ascertained. But a transcript made by or for Ghelen would eventually, after editorial correction by him, be indistinguishable from the final text of the extant printed book. What I am trying to show is, not that Ghelen did not at some time have his own working copy of the *spirensis*, but that he actually saw and used that book, and that his text is, beneath a layer of alterations, a direct witness to Σ independent of C M P and V.

43. See my critical note *ad loc.*, where CMPV's incredible *nec te* in *d.r.b.* praef.5 is also considered in this light. Could CMPV's *librarum* for *libyarum* at *notitia* or.i.40 have been in Σ? - that is, did Ghelen find more to change here than the number in conjuring up his emendation *Libyae*?

44. The copy which Bilde eventually dispatched to Froben consisted of two books, the 'monstrously corrupt' *murbachiensis* itself and an even fouler transcript hastily made from the Murbach text by a friend shortly after the exemplar had been brought to light. The printers were directed *ut uetus exemplar non minus quam exemplum inspicerent*: an instruction which might mean not only that the parent could be used to correct errors in its offspring but also that necessary corrections had been entered indiscriminately in both. (For a discussion of the transmission of Velleius, and of Bilde's editorial manoeuvres in producing his text, see Woodman's edition [1977] of Velleius 2.94-131, pp.3 *sqq.*). A notorious instance of an antique manuscript still extant which was used as printer's copy is the tenth-century *gandershemensis* of Hrotsuith, sent to Hoeltzel's shop in Nuremberg for Pickel's ("*Celtis*'") print of 1501 and never returned to Gandersheim; it survives, interpolated with editorial matter and smudged with printer's ink, in the Bayerische Staatsbibliothek (as *codex monacensis latinus* 14485). A possible case of a manuscript transcribed specifically to serve as printer's copy may be identified in a fifteenth-century Augustine *de ciuitate dei* now at S. Scolastica, Subiaco; this seems, from details of its readings and lineation, to have been the copy supplied for the *editio princeps* of the text, printed by Sweynheym and Pannartz at Subiaco in 1467. (Details and photographs of the Gandersheim and Subiaco codices in Adrian Wilson, *The making of the Nuremberg Chronicle*, Amsterdam 1976, pp.32 *sqq.*)

45. The collation is: roy.$2^0$, $*^8$ a-$o^6$ $p^4$ q-$r^6$ (on a problem in the last quire see next note), and the contents: *1r title, *1v blank, *2r-v Ghelen: *epistula nuncupatoria*, *3r-*5r Bilde von Rheinau: *Illyrici descriptio*, *5v Alciati: epistle prefatory to his essay *de magistratibus*, *6r-*8v Alciati: *de magistratibus ciuilibusque et militaribus officiis*, a1r-a3v *descriptio urbis Romae*, a4r-a6r *descriptio urbis Constantinopoleos*, a6v-p4v *notitia dignitatum*, q1r-r3v *de rebus bellicis*, r4r-r5r *altercatio Hadriani et Epicteti*, r5v colophon, r6r blank, r6v device.

46. I was once inclined to think that Froben had made a physical alteration in the make-up of the last quires of the printed book, since the leaf which should be r4 lacks a signature, contrary to the printer's practice in all the other ternios in the volume. Quire r could therefore, it seemed, be a duernio (in which case r4 would naturally be left unsigned), and the following leaves, the supposed r5 and r6, might not after all be conjugate with r2 and r1 respectively but might constitute a separate, unsigned cancel unio. This would imply that some change had been made in the contents at the end of the book, perhaps reflecting an abandoned intention to print *de gradibus cognationum* in this area. But examination of the water-marks and chain-lines of the paper seems to show quite clearly that 'r4', 'r5' and 'r6' are conjugate respectively with the signed leaves r3, r2 and r1; and the absence of a signature for r4 is probably the result of a careless (or aesthetically-inspired) omission of the direction-line below the picture-block which 'r4' carries, showing Hadrian and Epictetus in debate.

47. It is not necessary for me to say again that Ghelen did not use C (*supra*, §12) and cannot have seen this ascription there. But it may have existed as a note written into the *spirensis*, whence both C and Ghelen have reported it.

48. On purely temporal grounds, Moelbrigte's text of the *notitia*-group, if he ever had one, cannot have been the lightly-corrupted insular archetype of Σ, errors in which are tentatively isolated in my textual notes to *d.r.b.*pr.5 and xxi.1. To speak plainly, I am not inclined to believe that he imported a *notitia*-group manuscript at all. Nevertheless, the insular connexion with official and topographical texts is interesting and suggestive. Anglo-saxon foundations on the continent were stocked, presumably from their parents, with texts of this kind (for Speyer, see *supra*, note 25); the books - among them, I must insist, the *spirensis* of the *notitia* and its satellites - are or were in insular script; and we have proof that at least one eighth-century scholar, an Englishman working on the continent, had read one of the works in the *notitia*-group (a fact which may reflect unfavourably on the view that the group was unknown in continental monasteries until Moelbrigte's migration in the eleventh century). Alcuin writes (*epist.*88 *ad fin.*: Duemmler, *m.G.h. epp.* iv = *epp. aeui karol.* ii, p.133 lines 4-9: a.d.793-796?): *de epistola interrogasti, quid esset. nam* epi *super*,

stola *habitus grece dicitur. unde Hadrianus imperator Epitetum philosophum inter alias inquisitiones interrogauit quid esset cinctum. at ille uidens eum epistolam manu tenentem respondit* quod manu tenes, *uolens intellegere quasi supercinctorium esset epistolae sigillum, quo a foris uestiatur cartula. hanc habeto interpretationem, donec meliorem inuenias uel ueriorem.* The reference to the *altercatio Hadriani et Epicteti* is perfectly clear: I transcribe the relevant lines (they form the very opening of the text) from Ghelen's edition, sig. r4v:- HADRIANVS. Quid erit nobis si cinctum soluas, neq; nudaberis ipse? respice corpus, quo & doceri possis. EPIC. Epistola est.

49. See critical notes on *d.r.b.* pr.5 and xxi.1; note 43 above; and, for a possible feature of this manuscript's layout, the critical note on the chapter-heading of *d.r.b.*xvi.

## APPENDIX

## PREVIOUS EDITIONS OF THE TEXT

I give this conspectus substantially after Neher and Thompson. One preliminary observation may be of service to users of the Oxford edition. Thompson confesses (p.17 n.3) that he had not seen a copy of a 1729 reprint of Labbe's work mentioned by Neher. He therefore gives no bibliographical details to supplement Neher's bare references. Fortunately I can come to the rescue here. The curious reader will find our treatise execrably printed with the *notitia dignitatum* and its related texts by Javarina, in the folio *corpus byzantinae historiae*, Venice 1729, volume xviii, where *de rebus bellicis*, with no illustrations, occupies pp.32-36 of the second pagination series.

GHELEN (*Gelenius*): the *editio princeps*, with *notitia dignitatum*, &c., Basel (Froben) 1552. I transcribe the unwieldy title-page:-

> NOTITIA VTRAQVE/CVM ORIENTIS TVM OCCIDENTIS/ VLTRA ARCADII HONORIIQVE CAESARVM TEMPORA,/ illuſtre uetuſtatis monumentum, imò theſaurus prorſum incomparabilis./ PRAEcedit autē D.Andreæ Alciati libellus, De magiſtratib. ciuilibusq; ac/ militaribus officijs, partim ex hac ipſa Notitia, partim aliunde deſumptus./ CVI ſuccedit deſcriptio urbis Romæ, quæ ſub titulo Pub. Victoris circum/fertur: & altera urbis Conſtantinopolitanæ incerto autore, nunquā antehac/ typis excuſa, Imperialium uidelicet ac primariarū ſedium utriuſq; Reipub./ SVBiungitur Notitijs uetuſtus liber DE REBVS BELLICIS ad Theo/doſium Aug. & filios eius Arcadium atq; Honorium, ut uidetur, ſcri-/ptus, incerto autore. ITEM, ne quid de antiquo exemplari/ omitteretur, Diſputatio Adriani Aug./ & Epicteti philoſophi./ (*Froben's device*)/ BASILEAE, M D LII/ Cum gratia & priuilegio Imp. Maieſt. ad annos quinq;//

*de rebus bellicis* appears on sigs. q1*r* - r3*v*, with woodcut blocks of all the illustrations.

SCHRIJVER (*Scriverius*): in his edition of Vegetius, &c., Leiden (Plantin) 1606/7; *de rebus bellicis* on pp.81-102. Text from Ghelen with further emendations; picture-blocks (smaller than Ghelen's and heavily modernized) taken from passages of *de rebus bellicis* printed by Stewech in his commentary on Vegetius, Antwerp (Plantin) 1585, repeated twice at Leiden, separately in 1592 and with Schrijver's Vegetius in 1606/7.

PANCIROLI(*us*): a summary of the contents of *de rebus bellicis* appears in his edition of *notitia dignitatum*, &c., Venice (de Franceschi) 1593 and [posthumously] 1602. Panciroli never in fact edited the text of the treatise; but a new edition of his *notitia* work (Lyons [de la Porte] 1608) contains, in addition to the summary, the *de rebus bellicis* itself (on pp.201-222). A further edition (Geneva 1623) re-arranges the materials and relegates the treatise to an appendix (pp.24-39). It is not known by whose authority the text - Ghelen's, modified only by misprints - was inserted into the 1608 edition of Panciroli: *Rhuardesius* of Cahors, who added notes to Panciroli's commentary, or *Maranus* of Toulouse, who prepared the indices to the volume, may have been responsible for its inclusion.

Ph. LABBE: in his *notitia dignitatum*, &c., Paris 1651: *de rebus bellicis*, without illustrations, on pp.164-189. Ghelen's text, very slightly altered. Reprinted in Javarina's *corpus byzantinae historiae*, Venice 1729 (see my headnote *supra*). One of Labbe's readings is a great curiosity, and deserves to be recorded here, since it will find no place in an *apparatus uere criticus*. The title of Picture II, FELIX INCHOATIO *e.q.s.*, is tacked on to the end of c.iii by the addition of the words *ut sit*. The picture itself being suppressed, the resulting text has a certain specious plausibility: *formas et magnitudinem tam aereae quam aureae figurationis pictura praenuntiante subieci, ut sit felix inchoatio sacrae diuinaeque monetae*. The cue for this feat of misdirected ingenuity was no doubt taken from the closing sentence of c.vi, *imaginem tormentorum ... orationi subieci, ut sit facilis imitandi confectio*.

SCHNEIDER: in *anonymi de rebus bellicis liber*, Text und Erläuterungen; Berlin (Weidmann) 1908. A worthless *verbatim* reprint of Ghelen's edition, of interest only in that it reproduces, in a reduced size, the splendid picture-blocks of the *editio princeps* - an insufficient excuse, one feels, for the reproduction of the sixteenth-century text as well.

S. REINACH: in *Un homme d'idées au bas-empire* (16 *Revue archéologique* [sér.5], 1922, pp.205-265); text of *de rebus bellicis* on pp.255-265. Noteworthy as being the first attempt to control the Gelenian *textus receptus* by collation with a surviving MS (in this case *parisinus latinus* 9661, our P): adequate recognition of this in Thompson, p.17. The text which Reinach presents, however, is marred by faulty collation, and is so disfigured by misprints and bad readings as to belie even the modest claim (p.205) that it is *à peu près lisible:* adequate condemnation of this in Thompson, p.18.

E.A.THOMPSON: in *A Roman Reformer and Inventor: being a new text of the treatise* de rebus bellicis, *with a translation and introduction by E.A. Thompson*: Oxford (Clarendon Press) 1952. Text on pp.91-105, English paraphrase on pp.106-123. The publication of this book, precisely four hundred years after that of the *editio princeps*, marked an epoch in the study of the treatise: its editor rescued from oblivion a work of crucial importance for our study of the fourth century; he provided the first acceptably critical edition of the text; he took the unknown author seriously, and commented on his ideas with much learning; and he laid, single-handed, the foundations on which no less than fifteen specialists have built in the present volume. It is not surprising that, amidst such a diversity of materials, some parts of the work should seem less satisfactory than others, and that certain legitimate topics of investigation (*quae quidem non ignota sunt illi, quem alia plura sollicitant a nobis aliena*) should have been neglected. The text (it may be thought) required greater firmness of handling, the variants more attentive consideration, the style more subtle comment. I must not lay myself open further to the charge of envy or ingratitude, particularly since I am wholly indebted to Thompson's meticulous collations for my reports of the MSS: *uerum quia illum quaedam occupatum effugiunt, otio persuasus, non adeo a rerum commoditatibus peregrinus, utilia undique redacta conferre gestiui. si quid uero libentius oratio mea pro rerum necessitate protulerit, aestimo uenia protegendum, cum mihi promissionis implendae gratia subueniendum est propter philosophiae libertatem.*

EXCURSUS

# THE PRINTING OF THE *EDITIO PRINCEPS*

We noticed above (*Transmission of the text* §18) that Ghelen was already at work on preparations for the printing of the *notitia* volume in 1536: the gap of sixteen years between this date and Froben's eventual publication of the *editio princeps* in 1552 remains unexplained. Is it possible that the book was intended to appear earlier than this? that part of it was actually set up in type in 1536? I intend here to set out the facts concerning the dated plates - for *two* plates, not one, carry dates; to establish the consonance of these facts with the theory that Ghelen's exemplar was the *spirensis, nunc deperditus* itself; and to produce bibliographical evidence to show that, once the decision to print had been taken, the book as extant was set as a continuous whole. In no way can my observations be checked against modern reproductions of the texts and illustrations; the reader who wishes to verify them must consult an actual copy of the 1552 print.

## 1 THE DATED PLATES

The first date which we are given, .M.D.XXXVI, appears in large capitals on the *insignium* grasped firmly in the left hand of the majestic allegorical figure of *prouincia Palestina* on sig. h2v. The linked initials SG appear towards the lower right-hand corner of the plate; the monogram recurs in the plate of *Italia, Illyricum, Africa* on sig. l1r, on the rock in the centre foreground. The date 1536 consists well with Ghelen's known operations in the mid-1530s. In 1533 he had published his Ammianus, and in 1535 his castigations of the Elder Pliny and the Livy in which he collaborated with Bilde von Rheinau. These printed, he could turn his attention to the complex task of producing the elaborately-illustrated *notitia*-group, discovered by Bilde at Speyer as long ago as 1525 (above, *Transmission of the text* §18). Preparation of the plates must have begun directly, and *prouincia Palestina* at least was ready in 1536; but for some reason progress was suspended, and the book, which is wholly of one setting, cannot have been put together until 1550 at the earliest. The *extant* book, that is: I cannot of course preclude the possibility that a few pages of the text were in type in 1536, but if so, they must have been broken up when it was found that publication would have to be delayed. No sixteenth-century printer would keep an incomplete set of type-pages standing for sixteen years; the compositor's time spent in re-setting was less valuable than the ready availability of the type. The evidence that the plates were still being prepared in 1550 comes from an inscription on another picture-block, which appears first as the device of the eastern *magister scriniorum* on d4v and again as the device of the western *magister scriniorum* on l4r. (At its second appearance, I should add, the plate prints very coarsely, suggesting either that the block was soaked in ink from its previous use on d4v or that a particularly heavy pull was applied to the press - perhaps both; and in fact the plate, when it appears on l4r, has been seriously damaged, either by splitting of the block or by accidental gouging with a pointed instrument: three discontinuous white lines, about 2mm across, run vertically from top to bottom of the plate, about three-quarters of the way from the left edge). Towards the bottom of this plate, touching the left margin, is an open book, on which can be read the following Greek letters:-

| *left page* | *right page* |
|---|---|
| Ξ (or E) ȣ (?) ς (?) Φ | Π Α Ρ Α |
| Α Φ Ν (the N reversed) | Ι Ω Α Ν |
| | Ν Η Ι Ω |

To the right of and below the book, touching the lower margin, is a partly-unfurled scroll, on which appears:-

Α Σ  Β Α Σ Ι Λ Ε Ι Α
Ε Κ Θ Ε Σ Ι Σ  Κ Ε Φ
Α Λ Α Ι Ω Ν  Π Α Ρ

Of the letters in the upper line of the left page of the open book I can make no sense, indeed I cannot even guarantee my readings; but the group ΑΦΝ is surely the year 1550 in Greek alphabetic numbers. On the right page of the book an articulation παρὰ 'Ιωάννη<ι> ιω suggests itself, and this should give us the name of the block-cutter, Johann; the ΙΩ, its Ι not quite certain (the page is drawn with a fold, on which the letter comes) may be merely an unmeaning filler for the line derived from the beginning of the name. (A division παρὰ 'Ιωάννηι ω is possible but even less explicable). On the scroll we have αϲ βαϲίλεια ἔκθεϲιϲ κεφα-λαίων παρ, from which I can derive no coherent sense beyond some reference to imperial dispositions in chapters, the subject and mode of treatment of the *notitia*. The final ΠΑΡ here - ΠΑΠ is just possible - may be a filler taken from the opening letters on the right page of the book above. I would be glad to have a convincing interpretation of these mysterious inscriptions; but, whatever the uncertainties elsewhere, the Greek date ΑΦΝ = 1550 is not, I think, an optical illusion.

A possible implication of the order of appearance of these dates in the print needs to be worked out separately. The block dated 1550 precedes (on d4v) that dated 1536 (h2v): were the blocks therefore not cut in the order of their appearance in the finished book? But the 1550 block recurs on l4r; the plates could therefore have been cut in correct sequence, with the 1550 *magister scriniorum* block projected forwards to d4v during printing.

Preparation of the plates, then, begun by 1536, was still under way in 1550. Was this really a task which would take fourteen years to complete? I very much doubt it. Master Hans, who traced the pictures in the *spirensis* for Ott Heinrich, took only a year over the work (above, *Transmission of the text* §3 with document F in note 4); and even allowing for the greater complexity of wood-block cutting, fourteen years seems an unreasonable time to spend in the preparation of the plates - unless, that is, access to the original from which the plates were made was restricted and intermittent. If Ghelen (and, through Ghelen, his artist) had had continuous access to their manuscript, the delay would not have occurred; this applies whether the artist made his tracings (to glue to the raw blocks) directly from the original or from drawings made from it; in the latter case one simply assumes discontinuous access to the original while the pictures were being copied. The possibility that this original was the *spirensis* is again seriously raised; the Dean and Chapter of Speyer may have been very unwilling to allow Ghelen extended use of their manuscript if it was indeed in the poor condition which they alleged (*'das alt exemplar sey etwas blod'*) as a reason for refusing it to Ott Heinrich in 1548.

## 2 CONTINUITY OF PRINTING

It now becomes necessary to demonstrate that the *editio princeps* as we have it is the product of an uninterrupted piece of type-setting; to discount, that is, the possibility that portions of what we now see (including at least quire h, in which the 1536 *Palestina* plate occurs) were in type in 1536, the rest being added piecemeal as copy became available towards 1550 (the *terminus* for quires d and l, with their dated block) and into 1552, when the book came from the press. I shall demonstrate the continuous setting of the printed book to the limit of the evidence provided (a) by foul-case letters or substitute sorts, (b) by the repetition of picture-blocks, and (c) by blind impressions of type kept standing to fill out the chase at the bottom of short type-pages. These findings will not only serve the purpose of establishing the typographical unity of the *editio princeps*: they will cast interesting light on the methods of a sixteenth-century printer faced with so complicated a piece of work as the setting of the *notitia dignitatum*.

(a) letter peculiarities (o for o; broken 'long s')

On a number of occasions in the book a circular, unweighted figure zero is used in place of the spheroidal, weighted lower-case o of the magnificent canon roman used by Froben for the cross-headings of the text. Thus on a2r, Regio VI; b1r, Prouinciae Daciae; c4v, Sub dispositione uiri illustris, Magi- and Ponticae III; c6r, Sub dispositione uiri illustris, Comitis (the first four words and the comma perhaps standing type from c4v); d2r Sub dispositione uirorum illustrium Comi- (first two words, possibly first three letters of third, perhaps standing from c4v and c6r); k3v, Intra Africam cum uiro spectabili Co-; k6r, Rationales; o2v, Dux Prouinciae sequanici.; p3r, Corrector Apuliae & Calabriae.; q4v, Clipeocentrus. Ti-chodifrus.; and q5v, Expositio Currodrepani. The last two instances are in *de rebus bellicis*, all the others in the *notitia*. Apart from the identical position of the o in the *sub dispositione* headings on c4v, c6r and d2r, suggesting standing type, there is no discernible pattern here, and the substitution of o for o is either a simple matter of foul-case or an attempt by the compositor to eke out his supply of o-sorts in the large type; probably the latter, since a foul-case letter ought not to survive more than one distribution of the type. The recurrence of the fault from quire a to quire q - that is, almost throughout the book - implies a continuous process of setting and distribution by a compositor making shift with an inadequate fount. Less widespread, but equally noticeable and pointing to the same conclusion, is the occurrence on five pages of *notitia* text, in cross-heads set in the canon roman, of a 'long s' with its upper curve broken as the result of an accident to the kern: thus on a3r, Obeliſci V.; e4v, Vicarius dioceſeos Ponticae.; f2r, Dux Iſauriae.; g2v, Dux Meſopotamiae.; and k3r, Intra Africam cum uiro ſpectabili Co-. After this the defective sort is avoided or (more probably) was scrapped. The pattern suits a normal rhythm of setting and distribution: a3r (second quire, third sheet, outer forme) distributed well before e4v set (sixth quire, third sheet, outer forme); e4v distributed just before f2r set (seventh quire, second sheet, outer forme); f2r distributed just before g2v set (eighth quire, second sheet, inner forme); g2v distributed well before k3r set (eleventh quire, third sheet, outer forme). The irregularity of o for o spans the quires carrying the dated plates, while the broken long s locks quire d (first appearance of the 1550 block) into sequence with quire h (containing the block dated 1536).

## (b) THE REPEATED ILLUSTRATION-BLOCKS

The similarity of appearance between certain of the manuscript illustrations to the *notitia* induced the printer to make economies in the block-cutting by using the same blocks in more than one place and for theoretically different illustrations. The necessary differentiation was accomplished by typographical means, with new letter-press over the head of a re-used block or dropped into it. The deliberate re-use of blocks or parts of blocks in a manner consistent with normal imposition will suggest - what it does not of itself conclusively prove - a continuity of setting and impression. Froben's shop showed such skill in arranging the repetitions that only at one point is it obvious to the casual eye that a block is being used a second time. The scheme of the repetitions is this:

(i) the full-page block of the *insignia uirorum illustrium praefectorum praetorio per Illyricum* on b1v, re-used as the *insignia uiri illustris praefecti praetorio per Italias* on h6v, and again as the *insignia uiri illustris praefecti urbis Romae* on i2r;

(ii) the three-quarter block of the *insignia uiri illustris quaestoris* (east) on c5r, re-used as the *insignia uiri illustris quaestoris* (west) on k5r.

(iii) the three-quarter block of the *insignia uiri illustris comitis largitionum* (east; *sic*) on c5v, re-used as the *insignia uiri illustris comitis sacrarum largitionum* (west) on k5v.

(iv) the three-quarter block of the *insignia uiri illustris comitis priuatarum* (east) on c6v, re-used as *insignia uiri illustris comitis priuatarum* (west) on l1r.

(v) the full-page block for *sub dispositione uiri spectabilis castrensis* (east) on d2v, re-used for the *castrensis sacri palatii* (west) on l2v.

(vi) the three-quarter block for the *primicerius notariorum* (east) on d3v, re-used for the *primicerius notariorum* (west) on l3v.

(vii) the three-quarter block for the *magister scriniorum* (east) on d4v, re-used for the *magister scriniorum* (west) on l4r. On d4v *Memoriae epistolarum* & *libellorum Graecorum.* is dropped in in type at the head of the block, below the top margin; on l4r this is changed to *Memoriae Epistolarum* & *Libellorum.* (first word not from standing type).

(viii) The illustration for the *comes orientis* on e1v, taking up three-quarters of the page, is composed of four separate blocks: at the top, the *insignia* (clasped book on draped table next to *kalamarion*); below this, three horizontal strips each depicting five allegorical female figures. The four components are separated by lines of letterpress each naming, in order, the province represented by the figure in the section of the wood-block strip immediately beneath. All four of the blocks which compose the illustration are re-used - the first no less than six times - as follows: the top block (the *insignia*) on e2v, e3v, e5v, l6r, m1r and m1v; the first horizontal strip (e1v captions *Palaestina, Phoenice, Syria, Cyprus, Cilicia*) as the third strip on m1v (new captions *Aquitania secunda, Nouem populorum, Narbonensis prima, Narbonensis secunda, Lugdunensis secunda*); the second horizontal strip (e1v captions *Palaestina secunda, Palaestina salutaris, Phoenice libani, Eufratensis, Syria salutaris*) as the second strip on m1v (new captions *Belgica secunda, Alpium Maritimarum, Alpium Poeninarum, Maxima sequanorum, Aquitanica prima*); the third horizontal strip (e1v captions *Osrhoena, Mesopotamia, Cilicia secunda, Isauria, Arabia*) as the second strip on e4v (new captions *Cappadocia secunda, Helenopontus, Pontus Polemoniacus, Armenia prima, Armenia secunda*) and again as the first strip on m1v (captions *Viennensis, Lugdunensis, Germania prima, Germania secunda, Belgica prima*).

(ix) The two large horizontal strips each containing three female figures which (with the *insignia* block re-used from e1v) form the three-quarter-page illustration on e2v are each pressed into service on two further occasions: the first strip (e2v captions *Lybia superior, Lybia inferior, Thebais*) as the second strip on e5v (new captions *Rhodopa, Moesia secunda, Scythia*) and as the lowest strip on l6r (captions *Tripolitana, Mauritania Sitifensis, Mauritania Caesariensis*); and the second strip (e2v captions *Aegyptus, Archadia, Augustamnica*) as the first strip on e5v (new captions *Europa, Thracia, Haemimontus*) and as the strip below the re-used *insignia* block on m1r (captions *Betica, Lusitania, Gallaecia*).

(x) The full-page cut on g1r depicts, in stylized fashion, fortified towns with their names dropped in above them in type; in the upper register appear a scroll and a clasped book inscribed FL. INTALL COMORD PR, boxed-in from the rest of the plate and followed horizontally by two cities, captioned on g1r *Gallaba* and *Callinicum*. Below this is a row of four cities (*Dabana, Banasa, Syna Iudeorum, Oroba*); below this, four more cities (*Thillazamara, Mediana, Rasin, Circaesia*); and in the lowest register, in the right-hand corner of the plate, a single city (*Apatna*). This isolated city, with the lower angles and rule of the frame, is removed when the block reappears on n6v (the cities of the surviving registers now named *Flexo, Arrabona / Vindobona, Quadriburgio, Alanoua, Aequinoctiae / Ad Herculem, Gerolatae, Carnuntum, Quadrato*), on o1v (*Augustanis, Phebianis / Submuntorio, Vallato, Ripa prima, Cambiduno / Guntia, Foetibus, Teriolis, Quintanis*) and on o3v (*Grannona, Blabia /*

*Venetis, Osismiis, Mannatias, Aleto / Constantia, Rotomago, Abrincatis, Grannono*). Finally the single city and the frame are restored at the bottom of the block, and the full plate is printed on o6v with captions *Salecto, Tabernae / Vico Iulio, Nemetis, Alta ripa, Vuangionis / Mogontiaco, Bingio, Bodobrica, Confluentibus / Antonaco*.

(xi) The most immediately noticeable example of a repeated block comes at the facing pages h4v/h5r, where the same block serves for *Diuina Prouidentia* (accompanied by *Scientia rei militaris, Virtus, Auctoritas* and *Felicitas*) and for *Diuina Electio* (accompanied by *Vernus, Aestas, Autumnus* and *Hiems*), the captions in all cases being dropped in in type.

(xii) The block for *Dux Moesiae primae* on g6v, consisting of three registers (top, clasped book and scroll followed by schematic cities captioned, in type, *Flauiana, Pincum*: second, three cities, *Tricornium, Aureus mons, Viminatium*: third, three cities, *Laedenata, Cuppae, Adnouas*) recurs on n2v unchanged save for the captions, which become *Columnatensis, Vidensis / Inferioris, Fortensis, Muticitani / Audiensis, Caput cellensis, Augustensis*.

(xiii) The block captioned *Argentoratensis* on n2r is repeated with the caption *Olino* on o2v.

(xiv) Finally, the illustration for *Corrector Apuliae & Calabriae* on p3r - the fortifications captioned there *Prouincia Apulia & Calabria* - is repeated for the *Praeses Dalmatiae*, the fortifications now named *Prouincia Dalmatia*, on p4r.

Most of the blocks listed here required no adaptation for re-use other than changes in the type-captions; those of which portions were variously repeated were presumably sawn into sections after their first appearance complete. This, I think, is certainly true of the g1v block, which loses only the lower register which is subsequently restored; it is very probably true of the e1v block, although the full plate on e1v might represent the vertical assembly of separate horizontal strips designed from the first for re-use in other combinations later in the book. The horizontal unity of these strips is certain; the lines dividing them into compartments sometimes run through part of a figure, for example at the junction of the first and second compartments of the upper figure-strip on d6v. The horizontal borders of the e1v strips are certainly, the vertical borders probably, cut on the block; the borders of the g1v block are made up of type-metal rules and angles which often fail to join end-to-end accurately. When the compositor lifted this block and its borders from g1v he cut off the lowest register at a junction between rules in the side borders, leaving the lower angles and rules standing with the separated portion, and the upper part of the cut prints on n6v and o3v without any lower border at all. As re-assembled on o6v the block prints exactly as it did on g1v except that the metal rules (especially the lower left angle) are even further out of alignment than before.

Attentive consideration of the facts reported above will show

(1) that no block or portion of a block is re-used where a continuous imposition-pattern would forbid it. Thus (a) if the upper strip of the e1v block is to be used on e1v, e2v and e3v, there are only three other pages of this quire on which a continuous imposition will allow it to appear, namely e6v, e5v and e4v: it actually occurs on e5v. And (b) it was perfectly easy (though very daring) to arrange that a block should face itself at h4v/h5r: the pages are in different formes of different sheets.

(2) that the adaptation of the g1v block suggests that the lowest portion of this cut was removed, left standing, and replaced during a continuous process of printing.

(3) that the unusually coarse impression of the block on l4r, compared with its former appearance on d4v, though possibly caused by a heavy pull or by the block standing proud of the letterpress, may be the result of over-inking a block still wet from its former use at press.

(4) that the e1v strips re-appear in a group, though with vertical permutations, in quires l and (especially) m, tying these quires at least into a unity; and that other repetitions link quire b with

quires h and l, quires c and d with quires k and l, quire g with quires n and o, quire l with quire n and quire n with quire o.

(5) that these linked sequences run through the quires containing the dated plates and span the typographical irregularities, and that thus continuity of setting and impression is practically assured.

The most elusive and typographically interesting piece of evidence for this conclusion remains to be presented. This is derived from

(c) THE BLIND IMPRESSIONS OF STANDING TYPE.

In casting up the copy for the *notitia* portion of Ghelen's book the printer aimed to synchronize his largest picture-blocks with the appropriate text by placing, wherever he could, the block on a verso and the explanatory matter - often very short - on the facing recto, so that picture and text were available on one opening. Between quires d and l the arrangement of the printed text is essentially such that a full- or three-quarter-page block on the left-hand page faces a paragraph of letterpress on the right. To enable the recto page to be locked up properly the foot of the chase was often packed out with one, two, three, or as many as eleven lines of standing type justified to the full measure. These lines had been extracted from formes being broken up, and were inserted casually - some of them inverted - from direction-line level upward; they were of course not inked, but the face of the type bit into the surface of the paper, and in a raking light the impression can be more or less clearly seen. It is not possible for me to give a complete catalogue of these vestigial impressions; the traces are often visible enough to prove that a line was there without being sharp enough to identify which line it was. It will be sufficient if I prove, from two inter-locking units of blind impressions, that all the quires for which evidence is available - namely quires a - o, after which the pages are set virtually solid - represent a continuity of setting and distribution.

The largest single group of blind impressions is found on sig. m6r, where eleven lines taken from previous formes fill out the foot of a page carrying only 16 lines (including running-title and direction) of inked type. I read the blind impressions as follows (suppressing long s and digraphs):

Et quae de minore Laterculo emittuntur.
Duces Limitum infrascriptorum decem.
Vexillationes Comitatenses.
Pseudocomitatenses decem & octo.
In Gallijs.
T I N O P O L I T A N A E D E S C R I-
Praepositi Bastagarum.
III Ponticae
NIVM, TAM CIVILIVM QVAM MI-
Correctores.
[a signature (?) 14]

The first line here is used (probably as standing type) several times in the course of the text; its latest appearance was on g6r. The second line is from l6r, the third from k2r, the fourth from i6v, the fifth from k4v, the sixth from a4r, the seventh from k6v, the eighth from c4v, the ninth from h5v (not a6v, as the height of the comma shows), the tenth from l5v, and the eleventh almost certainly from quire l (not i). Line five, standing from a4r, had been used to pack out the chase for c2v, in combination with the line following it on a4r (these two lines in correct sequence but inverted) above three consecutive lines lifted from a6v. Very faint traces of the m6r impressions seem to occur on the preceding recto, m5r; and this, taken with the tenuous evidence of alternating irregularities in the running-titles (versos: NOTITIA PROVINCIARVM or NOTITIA PROVINCIARVM, rectos: IMPERII ROMANI or IMPERII R OMANI) suggests the use of skeleton formes. What is absolutely clear is that the re-use of lines from earlier formes to pack out later ones proves beyond doubt a strict continuity of setting, printing-off and distribution from (as m6r shows) at least quire a to quire m: and the sequence can be extended into quire o by means of the link with quire l shown in this group of blind impressions of lines from l2r/v at the foot of o1r:

| | |
|---|---|
| (l2v) | Castrensis sacri Palatij. |
| (l2r) | Comes domesticorum.Comes domesticorū |
| (l2r) | PEDITVM EQVITVM. |

Quire n is in any case directly linked with quire a by the presence at the foot of both n5r and n6r of

N I V M, T A M C I V I L I V M Q V A M M I-
N O T I T I A D I G N I T A T V M O M-

from a6v (not h5v, as the position of the comma - on the line on a6v, above it on h5v, in otherwise identical type - proves).

Quires a to o, therefore, were set, machined and distributed without interruption: we can have no reason for not extending this sequence back to the preliminary quire * and forward to the final quires p q and r: as we have seen in section (a) above, quires p and q are linked with quires a, b, c, d, k and o by the wrong-fount o for o. Reciprocally, the variation between D E R E B V S and DE REBVS in the verso running-titles from q2v to r2v (D E R E B V S in the inner forme at q2v, q3v, r1v, r2v: DE REBVS in the outer forme at q4v, q5v, q6v) confirms the suggestion that skeleton formes could have been used in the *notitia* quires (b - p).

In sum, then: three kinds of evidence prove that the *editio princeps* of the *notitia dignitatum* and its satellites is a typographical unity: once printing was under way, there was no interruption of the ordinary sequence of composition, machining and distribution. No portion of the extant book was put into type in 1536, when at least one plate was available; another plate was cut in 1550, and still the book - which should have taken no more than two months at press - could not be got out until 1552, when at last the editor was in possession of all the material which he proposed to print and the book could be set up and run off. Froben's editor Ghelen and his artist Johann (?) evidently experienced considerable difficulty in gaining access to their exemplar, which can have been available to them only at infrequent intervals over a period of at least fourteen years; and we know of only one manuscript of the *notitia*-group, carrying a very particular type of text, discovered by a Froben editor in a normally co-operative library, but a manuscript to which access was, perhaps for more than one reason, made exceptionally difficult in the second quarter of the sixteenth century: the *spirensis, nunc deperditus,* itself.

# 4 TEXTUAL CRITICISM

## SOURCES OF TEXT

C oxoniensis canonicianus lat. misc. 378 (January 1436: *Transmission of the text* §2)

M monacensis latinus 10291 (1542: *Transmission of the text* §3)

P parisinus latinus 9661 (>1443: *Transmission of the text* §5)

V uindobonensis 3103 nunc tridentinus (1484: *Transmission of the text* §4)

Σ The lost *codex spirensis*, archetype of C M P and V (*Transmission of the text* §§15 *sqq.*). I apply this *siglum* only to readings in which the agreement of C M P V is taken to guarantee the reading of the archetype: readings in which the four MSS are thought to agree by independently committing the same error, or in the discussion of which the use of the general *siglum* would prejudice the argument, are marked 'CMPV'.

Ghelen Siegmond Ghelen, *Sigismundus Gelenius* of Prague [1497-1554], editor of the first printed text of *de rebus bellicis* (with *notitia dignitatum* &c.), Basel (Froben and Bischoff) 1552 (*Transmission of the text* §§ 7-14, 17-22 and Excursus).

**Title** DE REBVS BELLICIS

It may be doubted whether the author gave the title to the work.

**praef.3** pro conscientia ueritatis ipse mihi poenam praemii loco, si promissio secus cesserit, posco, nec me laus aut <praemium> prosequatur cum abunde sit in hac parte indignationem pro audacia declinare.

nec *scripsi* ne Σ || praemium *suppleuit Maas*

(i) *nec*, in its common late sense *neque quidem*, seems to be the least change required to restore order and coherence to the structure. Limited emendation is clearly required in this area, despite Thompson's discouraging comment (p.24 n.2), but the text is not so corrupt as to need the extensive alterations which Maas (reported by Thompson, *loc. cit.*) was 'inclined to believe' desirable. The fault in the transmitted text - apart from the palpable loss of a word between *aut* and *prosequatur* - lies with *ne*, since *ne ... prosequatur* will not construe as a final clause running on from *posco*, and the alternative to emendation, a heavy stop before *ne*, results in a pause and an abrupt prohibition which fracture the argument and style of the period. I take the thought of the sentence, when reconstructed as printed above, to be this: 'so sure am I that I speak the truth that, if I am proved wrong, I insist on receiving punishment, not a reward; nor in any case am I to be rewarded or praised when it will be ample recompense to me, in a matter such as this, to escape the consequences of the emperor's wrath when he is informed of the truth by a private citizen.' Though this is by no means the most shapely or fully-evolved utterance ever devised by a Latin writer, the sense is acceptable and the expression continuous, and the phrasing is better adjusted to a context written in - for its author - a comparatively fluent, integrated and reasoned style. We still have to contend with two ellipses, the first accidental, the second apparently calculated. By *praemii loco* we must understand 'in place of the reward *which I might expect if my predictions prove accurate*': the *anonymus* surely did not expect a reward *si promissio secus cesserit*, although an earlier inventor had received a substantial payment even though his proposal was rejected by the emperor; Suetonius, *diuus Vespasianus* 18, *mechanico ... grandis columnas exigua inpensa perducturum in capitolium pollicenti, praemium pro commento non mediocre optulit, operam remisit, praefatus sineret se plebeculam pascere.* The second ellipse entirely depends on our reading *nec*: this is the ellipse of a statement of the contingency envisaged for the legitimate expectation of a reward, namely, 'if my predictions *do* prove accurate', which *nec*, with its implied *quidem*, would sufficiently suggest, and which the *anonymus*, with commendable modesty, declines to express openly. I do not think this subtlety wholly adventitious; it is at least in keeping with the self-deprecatory tone of the sentence.

(ii) A word has obviously dropped out of the text between *aut* and *prosequatur*. Maas' supplement *praemium* fulfils the demands of sense, style (for the repetition of words in the *anonymus* see my discussion in *Language and Style*, section III) and palaeography. For the idiom compare *codex theodosianus* 9.34.7.11-12 *laudem maximam ac praemium a nostra clementia consequetur*. Maas' alternative proposal, *honor*, is altogether less satisfactory here; but *theodosianus* 9.14.3.30 has *et praemio a nobis et honore donabitur*. Ghelen, faced with the incoherent *laus aut prosequatur*, silently deleted *aut* - a deadly expedient much favoured by him at similar moments of textual crisis elsewhere (see *Transmission of the text* §17).

**praef.5** qui recte quicquam sentire fuerint approbati:

recte *Maas* nec te Σ

Since it is exceptionally improbable that C M P and V would independently commit the misreading of insulars and the false articulation of letters evident in *nec te*, the reading may confidently be assumed to have stood in Σ. If the mistake goes back no further, i.e. if the reading *nec te* was produced by Σ's scribe, then this tells us something about Σ's exemplar; that it was in insulars without word-division. Further evidence for an insular pre-archetype (with a mild reservation on its use) is proposed in my note below on xxi.1 *comparatis*.

**praef.7** nam cum barbarae gentes neque facundia polleant ... minime tamen a rerum inuentione ... habentur alieni.

alieni Σ alienae *Ghelen editoresque deinceps*

*alieni*, in false concord with *gentes*, is *lectio difficilior* to the point of solecism, but I have allowed it to stand, supposing that we are dealing with either a *constructio ad sensum* or an extended use of the adjective whereby it signifies *homines alieni*. But the retention of Σ's reading may be an act of excessive piety towards the archetype, since the substitution of *i* for final *e* or ę occurred repeatedly in the Σ text: so, for example, *notitia dignitatum* or. 42.39 *macedonici* CMPV (write *macedonicae*), oc.11.67 *benetii* CMV *beneni* P (write *uenetiae*), oc.11.70 *tripolitani* CPV, omitted in M (write *tripolitanae* and compare the variants at oc.20.13). The *anonymus* certainly committed false concords at vii.6 *hoc ballistae genus ... radiis intorta iaculatur* and xviii.8 *uictricem exercitus cursum* (ix.1 *parma, hoc est modicus clipeus, fixoriis ... munitus, uarie habetur idoneus* is very different); but he was not so careless in xix.4, where correct agreement is preserved across a long hyperbaton: *Persarum sane gens, cui praeter ceteras nationes et dolus cordi est et corpori suppetit uirtus, quadratis est agminibus et maiori bellorum apparatu superanda.*

**praef.9** consequemini ex hoc opere commoda singulorum,

consequemini CM consequimini PV *Ghelen*

I have no doubt that *consequemini* is the better reading - compare the future tenses throughout xviii.7-11 - and that the author wrote it (which is not necessarily the same thing); but I am equally convinced that its appearance in C and M is not the result of superior accuracy of transcription. Even Ghelen, who was liable to alter anything on the slightest provocation or none, and who, I believe, knew Σ directly and PV not at all (see *Transmission of the text* §§12 ff.), would not have changed the future into the manifestly inferior present; the surprise is that he did not change the present into the expected future. We must accept, I think, that Ghelen gives here the unadulterated Σ reading, which P and V also reproduce - P V and Ghelen coincide in another crucial and apparently archetypal reading at ix.1, where see my note - and that C and M have stumbled upon the truth by accident. To retain the present and treat it as a future would be excessively harsh: the one clear instance of *praesens pro futuro* in the text (i.1 *non amplius bellorum florebit improbitas sed collatorum potius defecta subsidia recreantur*) is forced upon the author by the clausula: there is no such excuse for this rare licence in praef.9.

**praef.10** referemus enim quem ad modum, remissa tributorum medietate, in robur proprium prouinciarum cultor habeatur:

*totum hoc comma errantibus oculis praetermisit* M

**praef.10** unde etiam, <in> ratione exactionum cessante contumelia, limitum solitudinem erectis castrorum munitionibus incola securus illustret:

in *suppleui, cetera secutus ueterem interpunctionem*

The supplement allows us to preserve a punctuation which reflects the natural run of the words. Thompson, following Maas, reduces the text to incoherence by writing *ratione exactionum ***, cessante contumelia limitum, solitudinem, erectis c.q.s.* This is perverse: to split up *limitum* and *solitudinem* spoils the controlling rhythm and robs the *anonymus* of a good phrase without substituting a better. This exercise in re-punctuation is consequent upon Maas' supposition of a lacuna after *exactionum*, to supply which he proposes, for example, *deminuta*. But the halting construction of the received text can be cured by less wholesale means, and I am not sure that two consecutive ablatives absolute would not be unstylish in an author who elsewhere deploys the construction rarely and singly (as in praef.15) and who twice (in xx.1) seems deliberately to avoid committing himself to it (see *Language and Style* 1 a, under *Ablative*).

**praef.10** uel quo argumento extra solitam largitatem cumulatus honoribus miles exultet.

argumento MV argumenta P argento C

**praef.12** ita ut hae per eam sine auxilio cuiusquam turbae obruantur.

turbae *omissum in* C

**praef.13** talis est excogitata sollertia ut equus, siue aciem rupturus inuadat siue fugientibus ingerat se, tali arte muniatur ut semet ipsum uerberans sine cuiusquam magisterio efficiat magnas hostium strages.

equus *Reinach* eques Σ

Gregory of Tours could use *eques* for *equus* (*ascenso equite* [better *aequite*], *hist. Franc.* 4.13 [p.151.1 Arndt]) but I cannot persuade myself that the *anonymus* did.

**praef.15** quae quidem non ignota sunt proximis uestrae clementiae, quos alia plura sollicitant a nobis aliena.

nobis MPV uobis C: *falso Thompson autumat cum* C *Ghelen facere*

**praef.17** cum mihi promissionis implendae gratia subueniendum est propter philosophiae libertatem.

propter *Ghelen* prope Σ

**i.1** bellicam laudem et gloriam triumphorum utilitas semper intueatur aerarii, ne profusa largitio semina magis excitet proeliorum;

intueatur *scripsi* imitatur Σ comitatur *Schrijver: post* aerarii *lacunam indicauerunt Reinach, Thompson, cui supplendae ille* cauendum uero *proposuit*

I offer *intueatur* more *exempli gratia* than as the final solution of the difficulty: *scrutetur* or *rimatur* would give the desired sense but are not easily reconciled with the corruption; *inquiratur* is palaeographically feasible but otherwise unsatisfactory. If we are not to hazard a correction we must at least obelize *imitatur;* the 'utility' (or 'service' or 'institution' - the *anonymus* often uses *utilitas* with deplorable vagueness) of the treasury cannot be said to 'imitate' or 'resemble' anything. Schrijver's correction is not unattractive, but I would prefer it in the form *comitetur:* the *anonymus* is clearly making a recommendation for future policy, and a jussive or hortatory subjunctive is the only appropriate mood for the verb. I do not think, with Reinach, that there is a gap in the text after *aerarii:* his proposed supplement is in any case very poor, substituting a tautology for a consequence.

i.3 patrum nobis est paulisper prouidentia referenda quam in rebus egenis habuerint, uel quae nunc <ita> in tanta facultatum licentia uideatur.

ita *textui adieci*

Maas suggested *uideantur* (sc. *egenae*); that is, understanding the comparison to be with the relative poverty of antiquity; but I fail to see how this improves on the transmitted text, where *quae nunc* ... *uideatur* refers back to *prouidentia* and the comparison is with ancient frugality. In the heat of argument the *anonymus'* structures are often disorganized, and it is over-precise (a harsher word could have been used) to insist that *quae* must inevitably refer to the nearest preceding feminine. What the transmitted construction does cry out for is adverbial completion, and *ita* is the likeliest candidate for mechanical omission.

i.6 ut uero emendi et uendendi utilitas * * * et ut facultas regiae largitati suppeteret,

*post* utilitas *lacunam exhibui*

Both the argument and the syntax of the received text are highly suspect here, though the editors have voiced no dissatisfaction, and Thompson's facile paraphrase glosses over the difficulties. *ut facultas regiae largitati suppeteret* makes excellent sense and good Latin, but, with the text as it stands, *emendi et uendendi utilitas* must also be thought *regiae largitati suppetere,* and I do not understand how even the *anonymus* could have thought that it did. Nor can we accept the transmitted text and interpret *suppeteret* first absolutely and quasi-reflexively (in the sense of *cresceret*) with *emendi et uendendi utilitas* and then normally with *facultas regiae largitati*. The collocation *et ut* in any case prevents our referring *suppeteret* back to *emendi et uendendi utilitas* (though *et* alone would not have prevented our doing so) and this collocation, I think, inevitably shows that the text here originally contained two syntactically parallel clauses of which the first has lost at least its verb and perhaps an associated dative phrase as well. If *facultas regiae largitati suppetit,* then *emendi et uendendi utilitas* should benefit another class of interests in a parallel but distinct fashion, and *exempli gratia* one might restore *ut uero emendi et uendendi utilitas <populi commoditatibus inseruiret>, et ut facultas regiae largitati suppeteret, in aeris usum e.q.s.*

i.9 aeris autem materia, quae iam pro copia uilior erat,

pro] prae *Ghelen, gratuito*

ii.6 et per gradus criminum fouit tyrannos,

tyrannos *e* trrannos *effectum in* V

Confusion of *y* and *r* is the fault most characteristic of transcription from an insular hand. V's error reveals the script of its exemplar, and there can be no reasonable doubt that that exemplar was the *spirensis* (see *Transmission of the text* §4). That V is a direct copy of Σ, and not of an intermediary, appears from its correction *secundis curis* of the misread *y* here. C mistakes insular *y* for *r* at xix.3 *ascogefyri* (where see my note): and the script causes trouble again at vii.6 *intorta,* where diverging corruptions arise in M and V through misreadings of *t* and *r*: see my note there. For the insular parentage of P consult (briefly) my note on xii.3 *erigunt* and (more fully) *Transmission of the text* §5; for a proposed insular exemplar of Σ itself, my notes on praef.5 *recte* above and xxi.1 *comparatis* below; and for insular errors not exemplified in the text of *de rebus bellicis* but found elsewhere in the *notitia*-group, footnote 7 to *Transmission of the text* §5.

ii.6 tyrannos, quos ad gloriam uirtutis tuae produxit <inopia> magis quam succendit audacia.

inopia *ego,* fortuna *Maas exempli causa suppleuimus*

*quos ..., produxit magis quam succendit audacia* contains no intelligible opposition between the verbs, and a separate subject is needed for the first, as Maas (*apud* Thompson, p.5 n.2) realised, though his supplement, which starts too wildly from the context, fails to convince.

iii.1 inter damna rei publicae non ferenda,

non ferenda] conseruanda C *incuriose*

**iv.2** et tamquam sua rebus sufficere non possit iniquitas,

rebus *ut interpolatum damnat Thompson*

*rebus* is harsh only as being the first word of the group which intervenes between *sua* and *iniquitas*; syntactically it participates in the construction exactly as it did above in iii.1 *ne rebus possit interesse simplicitas*. The *anonymus* is echoing, as he so often does, a form of words which he has already used; see below, 5, *Language and style* III.2, *Repetition of words*, passim.

**v chapter heading** de releuando militari sumptu

releuando P (*itane?*), *Ghelen:* reuelando CMV

The agreement of CMV effectively guarantees the reading of Σ here, and Ghelen was critic enough to restore *releuando* in the title from *releuabit* in the text. For P's reading we are at the mercy of Reinach's and Thompson's silence, but the point is of no importance; if it really presents *releuando* here, then a curious reversal of the process seen in its *ante correctionem* corruption of *releuabit* to *reuelabit* in v.3 has produced the truth; but that Σ wrote *reuelando* there can be no reasonable doubt.

**v.4** hic quoque pari liberalitate otio donatus abscedat

otio *scripsi* alio Σ

I take the cue for my emendation from the phrase preceding in v.3, *honesta missione donatus ... otio gaudeat absolutus*: the *alium* following **otio* at four words' distance may have helped to cause the trouble. The close repetition of *otio* will be very much in the author's style (below, 5.III.2 *passim*), and for the expression, compare *codex theodosianus* 6.35.4.1-2 *qui ob spectatum laborem otio donati sunt*, 14.3.7.2 *otio et quiete donetur*, 12.1.89.4-5 *seque otio uel debilitate membrorum uel mentis torpore donantes*. Ghelen made a gallant attempt to preserve the received *alio*, accenting it adverbially with final grave and presumably taking it with *abscedat*; but this leaves *donatus* painfully exposed and (assuming that he meant his *aliò* to be local) produces tautology with the following clause (*uel ad alium ordinem ... accedat*). Thompson obelizes *alio* and records two hariolations, Barbara Flower's *alimonio* and Schneider's *aliorum*. The first - despite Thompson's approving *fort recte* - loses sight of the meaning of *alimonium* in a desperate quest for palaeographical resemblance between corruption and correction, and the second is completely unintelligible.

v.6 habitabunt limites, arabunt quae dudum defenderant loca,

limites, arabunt] limites Arabum, *Ghelen impudenter*

v.7 armis exerciti et minori [ut minori] utpote tirones stipendio subleuati,

ut minori *deleuit Ghelen*

Ghelen rightly excises *ut minori*, the result in (hardly before) Σ of a complex mechanical error, the scribe's eye wandering between *et* and *ut(pote)*.

v.8 et damnis non deerunt matura et parata in primis subsidia.

in primis, *minime gentium relegandum, eliminauit Thompson*

Deletion of *in primis* assumes scribal anticipation of the opening of c. vi, *in primis s(ciendum est...)*: but the phrase adds a desirable emphasis to *parata*, and the repetition of the idiom is in fact a strong argument for its retention as authentic in v.8 (below, *Language and Style* III 2, *Repetition of words, passim*).

**vii chapter heading** expositio ballistae quadrirotis

expositio *W.R.Smith apud Thompson:* exemplum Σ

*exemplum* anticipates the opening word of the chapter, *exemplum ballistae...* That the inevitable correction should have been delayed by four hundred years is either a measure of the critical inattention which has been paid to the text, or an oblique tribute to the authority of a Gelenian edition.

vii.6 sed hoc ballistae genus ... sagittas ex se non ut aliae funibus sed radiis intorta iaculatur.

intorta CP intoxta V in auram M, *omisit Ghelen qui et* eiaculatur *scripsit*

V mistook the long-tailed insular *r* for *x* (which in insulars may have its lower left limb considerably exaggerated), and M evolved *in auram* from a confusion of rounded *t* with the curve of *r* and a spurious division of the letters into words; a corruption which, given M's late date, perhaps indicates a growing unfamiliarity with insular hands. Ghelen's omission of the word is interesting: it is contended (above, *Transmission of the text* §§ 11-14, 17-20) that he had direct access to Σ in preparing his text; and if so, the unclarity in *intorta* which resulted in the errors of M and V may have been sufficiently serious to induce him to omit the word altogether and extend the verb by way of compensation. Had he been able to read *intorta* securely he would not, I judge, have been disturbed by the inconcinnity of *hoc ballistae genus ... intorta:* he would have left the root intact and emended to *intortum,* just as he changed *alieni* to *alienae* (following *barbarae gentes*) in pr.7. His extension of the verb is curious (it is certainly not needed *clausulae causa*); perhaps it is an example of the Gelenian *libido coniectandi,* assisted by guilt at the preceding excision.

viii.3 cuius axium extremitates et frons nec non et superior latitudo fuscinis et lanceis armatur

extremitate Σ *correxit Schrijver*

ix.1 munit inferius latitantes ab incursu comminus uenientum.

uenientum PV *Ghelen* uenientium CM

Acceptance of the theory that Ghelen used the *spirensis* directly in preparing his edition makes his evidence decisive for the reading of Σ here. The reduced genitive is required *clausulae causa:* the rhythm ´◦◦`◦´◦, *cómminus uèniéntum,* is found throughout the treatise (pr.1 *succéssibus cònualéscant,* 3 *audácia dèclináre,* 4 *láteat ìnquiréntem,* &c &c) but the extended version with final dactyl, ´◦◦`◦´◦◦, *cómminus uèniéntium,* nowhere. Vegetius once uses the reduced genitive of the participle in the same *cursus uelox,* and the manuscripts show predictable corruptions (3.4 *non pro inuidia suggerentum* [so AGΠ; *suggerentium* DVP *suggerendum* LMQ]).

**x picture caption** PLVMBATA ET TRIBOLATA PLVMBATA MAMILLATA CM
PLVMBATA ET TRIBVLATA PLVMBATA MAMILLATA P

*spatium uacuum picturae et titulis destinatum more suo reliquit* V: et *eiecit Ghelen*

**x chapter heading** expositio plumbatae et tribolatae CMV

*mero casu omissum in* P: et *eiecit Ghelen*

**xi chapter heading** expositio plumbatae mamillatae CMPV

I am willing - though Schrijver was not - to accept the variation between *tribolata* and *tribulata* in the spelling of the name of the first javelin (see my note on xi.1, below); but I cannot believe that the author applied differently *constructed* names to almost identical weapons. In the name of the second device, *plumbata mamillata*, the word *plumbata* has substantival force (as it does at Vegetius 1.17 *plumbatarum ... exercitio:* the lexica, overlooking the *anonymus*, record plural instances only): and why should it not in the first? In my note on the picture-caption and chapter-heading of c. xvi I argue, from the presence of an insular corruption in a capital inscription, that in the original state of the work headings were affixed only to the chapters of text, not to the pictures which precede them. In that case we would have an explanation of the presence of *et* in both the first picture-caption and the heading of c. x, and a clue to the reason for its intrusion. If, at some early stage of the transmission, the picture of the javelins bore no caption at all - the reader is invited to test the following speculation by masking the printed caption with a slip of paper - then a copyist, faced with the heading of c. x in what Ghelen and I suppose to have been its original form, *expositio plumbatae tribolatae*, might easily conclude that the words *plumbatae tribolatae* referred *separatim* to the two weapons in the picture; their terminations are after all identical, and *plumbata* as a substantive may have been familiar from elsewhere. The insertion of *et* would make this supposed distinction clear. The subsequent provision of captions for the pictures, by a more intelligent or more enterprising scribe, who realised that the *mamillata* as well as the *tribolata* was illustrated, led to the transference of the relevant portion of the first chapter-heading in its interpolated state. I therefore recommend acceptance of Ghelen's deletions in the caption and the heading. Restoration of parallelism in the captions and titles of cc. x and xi could admittedly be achieved by *supplying* <et> in the second pair (PLVMBATA <ET> MAMILLATA: *expositio plumbatae* <*et*> *mamillatae*) rather than by excising it from the first: but this seems altogether undesirable in view of the demonstrably substantival nature of *plumbata*.

**xi.1** similibus locis <ac> in tribulata plumbo et pennis adhaerentibus,

ac *suppleui, ut noster latinius uel certe latine loqui uideatur*

There are alternatives to my supplement. *tribulata* is the orthography of Σ, but I do not follow Schrijver in extending this spelling to the caption and heading of c. x, where the manuscripts indicate an *o* in the weak pre-tonic syllable.

xii picture caption — CVRRODREPANVS

*sic scribendum:* CVRRVS DREPANVS CM CVRRVS DRIPANVS P: *deest ut semper pictura cum titulo in* V

xii chapter heading — expositio currodrepani CM

expositio curri drepani PV

C and M present, in the chapter-heading, what must be assumed to be the authentic form of this strange name; the *anonymus* invariably and correctly inflects *currus* in the fourth declension, and *curri drepani* is a barbarous simplification. Even where the word re-appears in its correct literal form an impression seems to have persisted among scribes that its components should be separated: so P punctuates the picture-captions to cc. xiii and xiv CVRRO.DREPANVS.SINGVLARIS, CVRRO.DREPANVS.CLIPEATI (*sic:* see my note *ad loc.*) Σ's reading in the chapter-heading cannot be guaranteed, and I do not see that anything is to be argued from the variants for or against the theory that the picture-captions were extracted from the chapter-headings at an intermediate stage of the transmission.

xii.2 duo uiri uestitu et armis <e> ferro diligenter muniti

e *suppleui*

For my supplement compare exactly Modestus, *de uocabulis rei militaris* (ed. Stewech in Schrijver's *Vegetius* &c [Lippenheim 1670], p.374 lines 3-4), *aculei ingentes ex* [*e* Stewech's margin] *ferro*.

xii.3 funes ... laxati quidem explicant, repressi autem erigunt falces.

exigunt P (r written above the x, with a dot beneath it)

The variant is crucial for determining the parentage of P, and deserved a better fate than relegation to the limbo of Thompson's supplemental apparatus. P has misread an insular *r* as *x*, precisely as V did in *intorta* at vii.6. The corrected error, taken in conjunction with other evidence, points clearly to an insular exemplar, and in fact to Σ itself: see *Transmission of the text* §5.

xiv picture caption — CVRRODREPANVS CLIPEATVS CM

CVRRO.DREPANVS.CLIPEATI P

*deest* V

An interesting and perhaps important variant. Either P has simply anticipated the genitive inflexion of the chapter-heading, or we have here the unique survival of an error transmitted by Σ and corrected (consciously or otherwise) by C and M, resulting from the extraction - in this case, singularly maladroit - of the picture-caption from the chapter-heading. The former possibility is much the more likely, but in view of the tendency of other scraps of evidence (see my notes on the captions and headings of cc. xi and xvi) the second cannot be entirely dismissed.

xiv.3 a secundo uero curru hac diuersitate secernitur, quod in illo unus equus et pondus bellatoris subit †dum ipsum equitem portat et† occulto temone iugum trahente minus amplectitur.

*locus desperatus*

Thompson, recognizing that the text breaks down in the latter part of the sentence, deletes *dum ipsum equitem portat* as a gloss, and marks a lacuna after the excised words. He was right to diagnose serious corruption in this area, though we might have been spared the awkward jests which he cuts on the subject in his preface, p.15. His deletion of *dum ipsum equitem portat* may be right; against its excision one might urge (1) that as an expression it is not untypically flat, (2) that *pondus bellatoris subit* is inflated but not obscure enough to need explanation, (3) that the author occasionally glosses himself (ix.1 *parma, hoc est modicus clipeus:* xvii.2 *in cuius alueo uel capacitate*), and (4) that *dum* at least, which reads rather oddly as the opening of a gloss on *pondus bellatoris subit,* could be genuine. If lacunae exist they could be extensive, and if *dum ipsum equitem portat* is authorial, matter may have been lost before as well as after it. Assuming that this phrase is not interpolated and that the surrounding text is badly defective, Mark Hassall reconstructs the passage in English as follows:

*It is, moreover, distinguished from the second chariot by this point of difference, that there a single horse both takes the warrior's weight in carrying the rider and <draws the vehicle, being placed between two shafts; whereas this chariot> surrounds <its horses> to a lesser extent, since a hidden pole pulls the yoke.*

This he considers the minimum supplementation required to sustain what he takes to be the original argument; but recovery of the Latin is out of the question, and I therefore obelize at the outer limits of the textual disturbance.

xv picture caption, chapter heading, text

THORACOMACHVS

thoracomachi

thoracomachum .. thoracomachus .. thoracomachi ... thoracomacho

The manuscripts variously distort the name by falsifying or omitting the aspirates wherever it appears, though all the variants (noted by Thompson) point to the same original, and one must assume that the author could spell the word correctly. The true form is given by M and P in the picture-caption, M P and V in the chapter-heading, M in c. xv.1 and 3, and M and P at xv.4, where M corrupts the case-ending from ablative to nominative.

xv.4 hoc igitur, ut diximus, thoracomacho inducto

inducto *Ghelen* inducta Σ induto *Schrijver*

Thompson (text: *inducto* - apparatus: *scripsi*) is wrong in thinking his conjecture original.

xv.4 qui graeca appellatione ex tuitione corporis nomen assumpsit

appellatione M appellatio CPV

Thompson attributes M's reading to conjecture, rightly; the agreement of C P V effectively guarantees the reading of Σ, whose text was disfigured not only by the perversion of terminations (observe *inducta* for *inducto* immediately above, and see my note on praef.7 *alieni*) but also, infrequently, by their actual omission (*notitia dignitatum* or. 33.26 *neocesar* CV *neocaesar* PM, write *neocaesareae*).

xvi picture caption

ASCOGEFYRVS

*sic scribendum:* ASCOGEFRVS CMP, *deest* V

xvi chapter heading

expositio ascogefyri

ascogefri Σ *correxit Thompson*

The form restored in the chapter-heading by Thompson (anticipated by Schrijver in his over-corrected ASCOGEFVRAE) is unassailable; see the variants at xix.3, below. It must of course be extended to the picture-caption, and an interesting point of palaeography is thereby raised. The omission of *y* before *r* is fatally easy in insular minuscules; not so in capitals; but perhaps we owe the capital script of the captions to the taste of the fifteenth- and sixteenth-century copyists. That Σ was in this portion of its contents at least an insular manuscript is sufficiently demonstrated in *Transmission of the text* §16, with my notes on ii.6, vii.6, xii.3 and xix.3. But even supposing Σ's picture-caption here to have been in minuscule, it seems highly improbable that the surviving copies (or Σ itself, if it were copying an insular exemplar - see on praef.5 and xxi.1) should commit the same mechanical error twice over. The identical corruption of *ascogefyrus* and *ascogefyri* strongly suggests an early state of the text in which headings were prefixed, in minuscule, only to the chapters of text, the name of each device being subsequently taken up, sometimes in a corrupt state, as a caption to the appropriate picture. This theory should be considered in the light of the chapter-heading error reproduced in the picture-caption to c. x and the P variant in the caption to the picture for c. xiv: see my discussions there.

xvi.2 his igitur ut dictum est, diligenter sutis,

I remove the intrusive editorial comma after *igitur*, since *ut dictum est* is not purely parenthetical but refers back to *his*.

xvi.2 et ita, in formam pontis adsociata, partibus explicatur integritas.

I comma-off *in formam pontis adsociata* to express the syntactical reference of *ita* to *explicatur* and not to *adsociata*.

xviii.7 his igitur, inuicte imperator, machinarum quoque inuentionibus communitam potentiam inuicti duplicabis exercitus,

communitam *Ghelen* communite Σ

Again a faulty termination in Σ: see on xv.4 above, and note *ducum* for *ducem* at xix.1 below.

xviii.8 portatilis pontis peculiaris succurret inuentio.

portatilis *Ghelen, quod erit* ἅπαξ λεγόμενον: portanti Σ: *de* portandi *scribendo olim cogitabam* xix.1 conueniet ducem spatia ternis agminibus ordinare,

ducem *Thompson* ducum Σ

xix.3 militarem uiam leuibus portatum uehiculis ascogefyri comitetur auxilium,

ascogefyri *Thompson* ascogefrri C ascogefiri PV ascogephyri M ascogefri *Ghelen e titulo c.* xvi

C's corruption betrays the insular script of its exemplar and, by thus revealing the Σ spelling, provides our oldest evidence for the true form of the name. P and V apprehend the *y* correctly and render it by the phonetically equivalent *i*: M unashamedly grecizes (the *f* is certain).

xxi.1 diuina prouidentia, sacratissime imperator, domi forisque rei publicae praesidiis comparatis,

comparatis *Gundermann* componatis CMV componatis P compositis *Ghelen*

The significance of the corruption here is worked out more fully in *Transmission of the text* §19 *med.* The confusion of *n* and *r* is easy and frequent in transcription from insulars, and examples of it are found amongst the copies of Σ in the *notitia*-group of texts (which includes *de rebus bellicis: Transmission of the text* §15); so in *notitia dignitatum*, oc.5.106 *fontenses* M *fortenses* CP *recte* (*deest* V), oc.37.13 *armonicani* CV *armoricani* MP *recte*. But it is noticeable that this confusion is encountered more often in CMPV together than in individuals of the group; so in *notitia* or.35.30 *panthorum* CMPV for *parthorum*, or.39.17 *armigenorum* CMPV for *armigerorum*, oc.1.56 *arnorarii* CMPV for *annonarii*, oc.3.31 *senoria* CMPV for *senonia:* and so at *de rebus bellicis* praef.5 *nec te* CMPV for *recte*. This suggests that some at least of these corruptions existed in the Σ text, that Σ's scribe was himself capable on occasion of misreading the letter-forms of an insular pre-archetype. Σ's *nec te* at praef.5, and the corruption in the heading to c. xvi, should be carefully considered with the possibility of an insular pre-archetype in mind; though against the theory one could argue that confusion of *n* or *y* with *r* might occur more readily to a copyist unfamiliar with insulars than to one who practised the script as his current hand. Here, therefore, at *d.r.b.* xxi.1, *componatis* may with little hesitation be assigned to Σ, but an unclarity in the writing of the letters *-na-* left P dissatisfied (as witness his expuncts) and induced Ghelen to change just these letters in order to produce an intelligible reading.

# 5 LANGUAGE AND STYLE

# I GRAMMAR

## 1 NOUNS AND PRONOUNS

### a] the cases

*NOMINATIVE*

A relative clause in apposition to, not epexegetic of, the subject in viii.1 *tichodifrus quod est genus machinae*, amounting almost to an inversion, *genus machinae quod est tichodifrus*. I deliberately avoid the too-obvious epexegetic articulation *tichodifrus, quod est genus machinae, ex rei suae commoditate e.q.s.*, which weakens the relative clause into superfluity (no-one could be unaware of what the *tichodifrus* was); if that punctuation is maintained we have an instance of the perfectly normal prospective attraction of the gender of a relative pronoun by the dominant noun (...*beneuolentiam,* qui *est amicitiae fons*, Cicero, *de amicitia* 14.50).

*VOCATIVE*

The historical significance of the change of number from *sacratissimi principes* (praef.1) and *clementissimi principes* (praef.8) to *optime imperator* (ii.7) and *inuicte imperator* (xviii.7) is discussed elsewhere in this collection. The *captatio beneuolentiae* and compendious catalogue of the contents of the treatise seem to be addressed to the joint masters of the Roman world as an appeal of general interest to both; the body of the text, with its detailed proposals, is directed to the emperor (of the west?) whom it most nearly concerns. But in praef.4, hidden away between the plural vocatives of 1 and 8, there is a reference to a single *rei publicae praesul*. Is this mere carelessness? Is is a generalizing singular, 'any head of state'? Or is it a sign of revision, the author having originally addressed one emperor only, subsequently changing (for diplomatic reasons) singular to plural vocatives where direct address was obvious but overlooking the oblique reference in 4?

*ACCUSATIVE*

A probable accusative of respect in xiii.1 *nec distat quicquam a superiore machina*; but the sentence is not free from syntactical ambiguity. *machina* I take as ablative, with *quicquam* as accusative and the verb's subject understood from earlier in the sentence: but *quicquam* could be nominative, with *machina* ablative; or *machina* could be nominative, with *quicquam* as accusative of respect.

*GENITIVE*

A partitive genitive depends on a neuter plural adjective in the ornate, almost rhapsodic summary of the virtues of the *anonymus'* inventions in xviii.8 *per aperta camporum*, clearly as a device to heighten the tone. The construction persists: *quidquid opis* follows almost immediately (11). *nouitatis aliquid* is an effective way of arousing the reader's curiosity in xiv.1.

On the use of the genitive of the personal pronoun instead of a syntactically appropriate inflexion of the possessive pronoun see [b] below.

*DATIVE*

The *anonymus* shows some fondness for the **predicative** dative construction; so with the colourless *esse* at ii.7 *erit curae* and at xix.4 *cordi est*, and with a verb more expressive of a physical state at xvi.5 *impedimento consistat*.

*ABLATIVE*

A wide range of uses, none of them specially remarkable. Ablatives of **respect** with adjectives (xvi.1 *et compendio facile et usu praecipuum*) and with a verb (i.4 *non otiosis opibus sed condendis potius moenibus laetabantur*); of **separation** (ii.8 *abstinere opibus*, viii.3 *uacantia defensore loca*) and of **complement** (v.5 *expensis grauidam rem publicam*); the **modal** ablative (xvi.2 *uitulinis pellibus arabica arte mollitis*); **descriptive** ablative (xv.1 *thoracomachum mira utilitate ad leuamen corporis*); **instrumental** ablative (xvi.2 *aquam de puteis follibus hauriendam*); and ablative of the **agent** with and without preposition (xiii.1 *duobus equis tractus a duobus etiam bellatoribus gubernatur*); standard correlative ablatives of **measure of difference** in xviii.11 *quanto magis ... tanto praestantior*. In xviii.9 *fluctibus transcurrens* the ablative replaces an expected accusative. Some difficulties of classification: at v.6 *regiis donis opulentos*, is the ablative one of complement or of respect? I think the former. At vii.2 *omni latere in hostem sagittas impellat*, is *latere* ablative of separation (harshly used - we should rather have expected a prepositional phrase) or a curious variety of the local ablative? Again, I think, the former. The same problem, and the same answer, at xvi.2 *una parte circulis exstantibus*. In xvi.3, is *obliquo fluuio* accompanimental or local? and in xviii.2 is *duabus rotis* local or instrumental? In xvi.2 *uitulinis pellibus ... mollitis, ... his igitur ... diligenter sutis, fiunt utres in magnitudinem e.q.s.* the first ablative clause is presumably absolute and the second goes closely with the verb; although the second may itself be absolute, and in either case the first might be merely prospective. The **ablative absolute** is in fact more conspicuous by its absence than by its presence (it occurs, e.g., at i.8 *aeris redundante materia* and xi.1 *plumbo et pennis adhaerentibus*); a conscious attempt to avoid it seems to be made at two points in xx.1, *millenis interiecta passibus* and *circumdata quodam praesidii cingulo*, where the context should be very carefully considered.

## b] special uses of the pronoun

α The author four times affects the genitive of the reflexive personal pronoun instead of an appropriate inflexion of the possessive: ix.1 *ad soliditatem sui*, xi.1 *in extremitate sui*, xvii.1 and 3 *pro magnitudine sui ... pro mole sui*. This could well be interpreted as evidence for a Late Imperial dating of the prose; the use of the construction seems to increase steadily from the Silver Age (effectively from Seneca) onwards, and has become pervasive by the period of latest antiquity (e.g. [Justinian] *instituta* 3.19. 2 *fin.*: *natura sui dominio nostro exempta in obligationem deduci nullo modo possunt.*)

β The reflexive is once intensified, solely for purposes of euphony, at xvii.3 *proque machinis in semet operantibus.*

## 2 ADJECTIVES

### a] morphology

The ablative singulars of *prior, minor* and *maior* are written *priori* (xiii.1), *minori* (v.7) and *maiori* (xix.4), but the form is quite as frequent in early as in late writing.

### b] concord

False concord in xviii.8 *uictricem exercitus cursum,* where the gender of the adjective has been attracted by the adjacent noun; and again in vii.6 *hoc ballistae genus...sagittas ex se non ut aliae funibus sed radiis intorta iaculatur,* though here we can extract *ballista* from *ballistae* as subject of the final verb and participle, the transition smoothed by the intervening *aliae.* A less certain example in praef.7 *barbarae gentes...habentur alieni,* where *alieni* was the reading of the lost Speyer archetype. I have allowed it to stand - see my critical note *ad loc.* - although editors since 1552 have unanimously printed Ghelen's *alienae.* The adjectival agreement at ix.1 *parma, hoc est modicus clipeus, fixoriis minutis...diligenter munitus* is entirely natural.

### c] adjective as noun

Infrequently used by the *anonymus,* but see praef.16 *utilia,* iv.5 *casti et integritatis cupidi,* xviii.8 *per aperta camporum.* In the case of the neuters at least this infrequency of adjectives substantivally used is a consequence of the author's extensive command of abstract nouns.

### d] adjective for adverb

The adjective in viii.1 *quo protectior eadem ballista operetur* is most unexpected, but it is quite in keeping with the *anonymus'* general cast of thought and expression; the adjectival, not the adverbial, qualification shows that the substantival element in the clause preponderates over the verbal (see below, *Abstraction and substantival style,* in section II (*Vocabulary and composition*) 2). Compare the less arresting *ne constrictum latus liberior hostis infestet* in xix.8.

### e] adjectival degree

There may be some residue of comparative force in the adjective at viii.1 *facilior ... ascensus,* which could imply 'an easier means of ascent (*sc.* than has hitherto been possible)'; although a positive sense suffices (and a superlative cannot be precluded).

Heightening of adjectival and adverbial degree is, of course, widespread in late prose as a symptom of linguistic inflation; it is a conscious artistic device only in Ammianus, where comparatives used regularly for positives contribute to the sense of exaggerated realism which characterizes his style. And so, in the *anonymus*, *facilior* at viii.1 is probably no more than a reflection of routine late practice. But *tichodifrus ... non altior sed humilior fabricatur*, following at viii.2, both is and is not a typical instance of its type. There is no difficulty in understanding the adjectives as the positives which they surely are, and yet they have, I believe, been dragged into the comparative degree by the *facilior* just preceding. (The persistence of word, form and phrase in the *anonymus* will be considered in a later section). Observe, however, the genuinely comparative force of the phrase when it is repeated in adverbial form at xviii.3 *ad dirigenda seu altius seu humilius tela*, where the adverbs are intimately connected with the following *pro ut uocet utilitas*, much less closely with *nunc erigit nunc deponit*. For the rest, a superlative force may be perceptible in xix.3 *latior ... fluminum occursus*. xix.8 *angustiora loca* is a very doubtful case, but I suspect that the comparison may be real.

## 3 VERBS

### a] tense

Present for future in *recreantur* at i.1, evidently *clausulae causa*: alternatively, the present might be felt to have potential rather than future force, in which case *recreentur* would be an easy alteration. But I would not recommend substitution either of this or of Reinach's *recreabuntur* for the transmitted reading; *florebit* gives enough of a temporal hint to *recreantur*, and, so far as the restoration of a subjunctive is concerned, a future indicative replaces a permissive or hortatory subjunctive at xi.1 *uirga accipiet in se*.

### b] mood

Indicative after *constat quod* at praef.6, *sciendum est quod* at vii.6, and *admonendi sumus quod* at xvi.4; but subjunctive follows *sciendum est quod* at vi.1. (On this form of the oblique statement see [f] below). The indicative after *sunt quae* at vi.3 is unexpected. *pro ut* takes an indicative at xviii.3 but potential subjunctive at xix.2. More alarming is the indicative after causal *cum* at praef.17 *aestimo uenia protegendum, cum mihi ... subueniendum est*: the construction is sporadically attested for classical writing, but is generally associated with a much later period; so Vegetius 2.5 *nam imperator cum augusti nomen accepit, tamquam praesenti et corporali deo fidelis est praestanda deuotio*, and Gregory of Tours *de uirtutibus s. Martini* 1 pr. *sub finem* (p.589.14 sqq Krusch) *sed quid timeo rusticitatem meam, cum dominus redemptor et deus noster ad distruendam mundanae sapientiae uanitatem non oratores sed piscatores, nec philosophos sed rusticos praelegit?* 'Iterative' *cum*, however, if one is correct in identifying it at xix.1 *cum bellicus apparatus prouocet certamina*, takes the usual subjunctive.

### c] voice

α *uestire* appears in middle voice at xv.2 *ut ... membra uestientis labori sufficiant* and 3 *ne ... thoracomachus ... adficiat uestientem*. Not in middle voice, but used absolutely with a certain abruptness, is *gestientem* at ix.1.

β The passive infinitive in 1.9 *formas et species ... uariasque ueterum prouisiones exprimi colorum qualitatibus properaui* is very harsh; we would rather have expected the subject of the finite verb to persist into the dependent clause, whose accusatives would be the direct objects of the active *exprimere*.

### d] morphology

α Notice the double compounds *superinducere* (xv.3) and *superimpositis* (viii.2; but *superposita* later at xv.4). The first is post-augustan (Quintilian, Pliny), the second in Livy (38.56.3 &c) but perhaps best known from Statius, *siluae* 1.1.1.

β A termination is syncopated *clausulae causa* at ix.1 *ab incursu comminus uenientum*. The contraction is familiar in verse of all periods; it enters prose probably with Nepos (*s.u.l.* at *Thrasybulus* 4.2 *sapientum*), and will be found in Seneca (*sapientum* at *ad Polybium de cons.* 14.1 and [emended out by C.F.W.Mueller] at *ad Heluiam de cons.* 12.1): it is established in Tacitus (*ann.* 4.12.1 *dolentum*, 4.41.2 *salutantum* [but *salutantum* at 11.22.1 is rejected by the editors], 6.50.4 *gratantum*). It then apparently undergoes a period of disfavour, to be revived (in all conjugations, and not necessarily in clausula) in the fourth century (Nazarius *panegyricus Constantino* 26.2 *cadentum*; historia augusta *Auidius Cassius* 4.6 *agentum*; Ammianus 19.8.5 *squalentum*, 19.8.10 *uiantum*; [Aurelius Victor] *de uiris illustribus* 34.9 *potentum*, a widely-used form), and it becomes distinctly common in the fifth (Vegetius 3.4 *suggerentum*, Augustine *de ciuitate dei* 4.2 *bellantum*; frequently in Sidonius [*epist.* 1.7.3 and 10 *occurrentum*, *sancientum*], Alcimus Avitus [*epist.* 36 (p.66.7 Peiper) *munerantum*] and Ennodius [*uita Epiphanii* 108 *audientum*]). *uenientum* at *d.r.b.* ix.1 may therefore be used to corroborate a late fourth-century date for the treatise.

γ The inchoative form *capescit* in xvii.3 is of some interest, if this is genuinely the author's spelling. *capesco* for *capesso* presumably arose through false analogy with the large family of inchoatives in *-esso*, a confusion perhaps assisted by pronunciation (the palatalization of *c* before *i* and *e* in terminations). The form is distinctively late: it is found, for example, at Prosper of Aquitaine *carmen de ingratis* 779 (*capescere*: but *capessi* at 509), Petrus Chrysologus *sermo* 42 (Migne 52.318B; *capescit*) and perhaps Gregory of Tours, *hist. Franc.* 9.25 (p.381.26 Arndt; *ad regionem ipsam capescendam* [*s.u.l.*: variants *capiendam* and (most MSS) *capessendam*]). The *glossae Philoxeni* warn that *capessitur* is spelt *non per sc* (at C30 Pirie-Lindsay). Since *capesciit* in Tacitus *ann.* 12.30.2 and *capescere* in Appuleius *apol.* 86 are exploded corruptions, a consequence of accepting a date for *de rebus bellicis* in the third quarter of the fourth century would be to make the *anonymus*'s *capescit*, if this is really what he wrote, the earliest recorded instance (to my knowledge) of the form *per se*. As for the idiom, *pugnam capessere* is not uncommon in Livy (2.6.8, 2.20.10, 10.5.4, 10.19.15, 40.33.6).

For the implications of the frequentative *latitantes* see section [e] immediately below.

## e] participles as nouns

The *anonymus* is noticeably fond of qualifying these with an adverb: so viii.2 *latenter incedentium*, ix.1 *inferius latitantes* (the frequentative is sensitively used, expressing the continuous state of the verb's action), xi.1 *similiter obstantia*. In this last case, where the adjectival or substantival *similia* was an easy alternative, the author displays a certain sense of linguistic *finesse* in remembering the verbal origin of his substantive. He had perhaps displayed it once before in a different handling of the participle as noun, at praef.11, where *inuenta*, rather than *inuentiones*, is used to mean 'inventions', that is, actual mechanical devices: *inuentio* should strictly signify the abstract process of invention, and it is used in precisely that sense earlier in the preface (11) and at xviii.4: while *inuenta* denotes the result of practical activity. Yet the distinction is less obviously preserved in the phrase found at praef.14 and xviii.8, *pontis inuentio*, where *inuentio* verges on the concrete; and it breaks down altogether at xviii.7, where *machinarum inuentiones* is an exact equivalent of the *machinarum inuenta* at praef.11. Perhaps after all we should not credit the *anonymus* with too much subtlety. Certainly he stretches *inuentio* into the almost concrete sense ' design' at viii.1 *fabrica ... utili et commoda inuentione praeparanda* (a very quaint phrase); and both *inuentio* and *inuentum* mean something very close to 'discovery' at praef.7 and i.7 respectively.

## f] syntax of the verb

α Consecutive *quod* with a finite verb (indifferently indicative or subjunctive without distinction of meaning) replaces accusative and infinitive after verbs *sentiendi ac declarandi* at praef.6 *constat... quod... consecuta est*, vi.1 *sciendum est quod ... perstringat ... et ... appetat*, vii.6 *sciendum est ... quod ... iaculatur*, and xvi.4 *admonendi ... sumus quod ... sunt ... substernenda*. Not a grecism (whatever may have been the origin of the construction) but a colloquialism which invades formal prose relatively late (effectively from the second and third centuries, especially in early christian writing) and even then carefully avoided by fastidious writers. The point does not need to be laboured here; it is sufficient to hint that we are dealing with a late writer evidently unpractised in literary composition.

β The active infinitive appears as complement to *est*, in the sense of *debet* with passive infinitive, at praef.6 *est considerare*. There is no need to see a Greek construction here; Greek may have helped in the formation of the idiom, but it is perfectly acceptable as a Latin turn from the silver age; the example of Tertullian (as at *de pudicitia* 7.14 *est ... retractare*) may account for its vogue amongst christian writers. In Ammianus 30.4.8 *uidere est*, where the auxiliary has the force of *licet* rather than *debet*, the construction more nearly approaches ἔστιν ἰδεῖν, and is probably a genuine grecism, although the influence of contemporary practice, and of Ammianus' beloved Gellius (*noctes atticae* 13.25.31 *est ... inuenire*) is not to be under-estimated here.

γ For what the information is worth, the prolative infinitive at x.1 *hoc iaculi genus ... non ... consueuit emitti* finds a curiously exact parallel in Vegetius 3.4 *medicina ... consueuit apponi*.

δ *habere* in its extended sense 'to provide' is found at v.6 *huius modi ... prouisionis utilitas habet in augmentum prouincialium ueteranos* and xviii.10 *quae ... habet de nauigii facilitate remedium*. The required sense is in itself not wholly unclassical, but both usages are slightly unconventional. In the first case, *habet ueteranos*, the *personal* object of *habet* meaning 'provide, produce' is apparently unexampled elsewhere. And in the second case, *habet remedium*, where *habet* is used as in the Vulgate, *Tobias* 6.7 *quod remedium habebunt ista*, we seem to have moved away from classical usage in that the phrase ought strictly to mean either ' it is in possession of a remedy' (so *pantheras ... remedium quoddam habere quo cum essent usae non morerentur*, Cicero *n.d.*2.50.126) or 'there is a remedy for it' (so *habet curationem*, Celsus 4.2.5, and *habet curationes*, Vitruvius 1.6.3); neither of these senses is possible here. In Pliny *n.h.* 11.41.118 *uenenum hoc remedia secum habet* the reflexive makes an essential difference.

ε There is a painful and slightly bewildering passage in ii.8, where *considerare* simultaneously governs two distinct constructions, firstly a straightforward accusative of the inner object and secondly an indirect question: *famosa regna considera, ... qua haec honoris laude ... frugalitas ... commendet*. The oblique clause is itself very awkwardly put together, and the author's emotion has evidently triumphed over what little ability he possessed to express himself with directness and clarity.

## g] gerundive as future participle passive

A device favoured by later prose, prominent in the *anonymus*: praef.15 *asserens ... cunctam rem publicam praedictis remediis subleuandam*: praef.17 *si quid ... liberius oratio ... protulerit, aestimo uenia protegendum*: xix.6 *protegendi sunt ab omni telorum nube*. Elsewhere the strict gerundival sense is maintained: vi.4 *huius modi nationes ... armorum sunt machinis prosequendae*: xviii.8 *siue terra sint subeunda discrimina*. But such instances as viii.1 *erit ... huius ... compositio uel fabrica ... praeparanda* and xix.7 *falcatis curribus exercenda uictoria est* are distressingly ambiguous.

### h] ellipse of the verb

The auxiliary infinitive is omitted in indirect statement on three occasions in the high-flown preface, evidently as an elegance of style: 11 *adnectenda credidimus,* 15 *asserens ... rem publicam ... subleuandam,* 17 *aestimo uenia protegendum.* A finite auxiliary is omitted at ii.4 *domus ... clariores effectae* (the *repletae* earlier in the sentence is purely participial): and copulative *est* is dropped, to great rhetorical effect, in xviii.4 *hoc tamen mirae uirtutis argumentum.* At vii.1, *exemplum ballistae, cuius fabricam ante oculos positam subtilis pictura testatur,* a startling ellipse deprives the sentence of its verb, but the sense remains quite clear, and we need not puzzle over what must be mentally supplied.

## 4 ADVERBS

### a] forms and meanings

α The compound *desuper* is worthy of passing attention at xvi.2 and 3, *desuper autem una parte circulis exstantibus ... in marginibus autem firmitatis gratia desuper extensis.* The very close repetition (to be discussed later) is more interesting than the form, which is uncommon but classical.

β *parumper* of space, however, in x.2 *fistula eiusdem ferri parumper extensa,* is altogether exceptional. Gregory of Tours uses the adverb substantivally of quantity (*hist. Franc.* 7.45 [p.322.7 Arndt] *admiscentes parumper farinae,* and *de uirtutibus s. Martini* 2.32 [p.621.12 Krusch] *parumper olei elicuit*); and both his and the *anonymus*'s usages must depend on the replacement by *parumper* of *paulum* (as substantive in Gregory, as adverb in the *anonymus*). For the *anonymus*'s usage I can find no satisfactory parallel. He uses *paulisper* conventionally (i.3, ii.8); is his spatial *parumper* a personal peculiarity?

### b] adverbial degree

Comparatives with positive force at xiv.4 *falces ... longius ab axe porrectas* and (far less noticeably) ix.1 *inferius latitantes*; with positive or superlative force at vii.4 *deponitur celerius*; perhaps genuinely comparative at xi.1 *ut plumbi pondere et pennarum celeritate adiuta rotunditas teli facilius clipeos ... ualeat penetrare.* I take the comparative to have superlative force in the treacherous phrase *in tantum longius uadit* at xviii.5, though here again a positive fits the construction. Uncertainties in the correct understanding of adverbial degree are very similar to those encountered in dealing with heightened adjectives, on which consult 2[e] *supra.*

## 5 CONJUNCTIONS, DISJUNCTIONS, PARTICLES

### a] alternatives and possibilities

These are usually asymmetrically expressed. The negative alternatives 'neither...nor' are given in praef.7 as *neque ... aut*, and the quadruple denials 'neither...neither...neither...nor' are expressed in praef.6 by *nec ... nec ... aut ... aut*. The quadruple possibilities 'whether...whether...whether...or whether' appear in praef.9 as *seu ... seu ... uel ... siue*: the variation is conscious and deliberate. A much more hazardous use of words can be seen at xviii.8 *siue ... siue ... aut si*, where alternatives have replaced disjunctives; the three clauses are meant to denote parallel but independent eventualities, and we must take them as if they were introduced by *si ... aut si ... aut si*.

b] connectives

α The aggressive initial *ergo* reinforced with a near demonstrative at iii.2 *ergo huic quoque parti* and viii.2 *ergo hic idem tichodifrus* may be an author's mannerism.

β The *anonymus* shows a decided tendency towards close connexion of his sentences with the help of the usual apparatus of particles and conjunctions, *quidem, autem, enim, uero, etiam, ergo, igitur*. This seeking after connexion is often obtrusive (as throughout chapter xix, for example) and sometimes leads to redundancy (the *enim* at the opening of ii.2). The first sentence of each paragraph naturally tends to stand alone; but each succeeding sentence is coupled to its predecessor, whether or not any logical relationship exists, by some word of connexion or disjunction (the particles of c.xviii may profitably be studied from this point of view); and the impression given is of a writer relatively unversed in literary expression attempting an elaborate style of which an important feature is conceived to be the unremitting use of connectives to articulate individual sentences with respect to each other and to integrate them into the finished paragraph. But sometimes the transitions are so unskilfully managed that one inclines to suspect that the author is simply trying to hold a disparate set of thoughts together by the cheapest possible means. The *anonymus* is, in fact, least happy in his handling of conjunctions implying direct consequence; *quam ob rem* (i.3), *igitur* (iv.1), *nam* (xvi.2) form exceptionally harsh transitions in their contexts (while, conversely, *sed* in xvi.2 retains scarcely any oppositional force at all). The heavily-laboured preface is, however, organized with conspicuous success; the resourceful subtlety of articulation here should be contrasted with the awkward transitions of, say, chapter i.

## 6 PREPOSITIONS

### a] a, ab

Notice *a superioribus curribus ... dissimilis* at xiv.1, followed by *priori ... currui ... dissimilis* in 2. The dative, though never so common as a genitive, is very ciceronian, the *a* construction far otherwise: the earliest instances seem to be in Solinus (53.17) and [Lactantius] *mort. persec.* 8.1, and it becomes firmly established in later writing (*uulgata, Daniel* 7.19, Augustine *quaest. hept.* 3.45, Salvian *gub. dei* 6.1.4 [p.67.24-5 Halm], Sidonius *ep.* 4.24.3 [p.75. 7 Luetjohann], Cassiodorus *in ps.* 34.11).

b] de

*de* is found to replace other prepositions in accordance with informal late practice: so for *e* at v.4 *miles de sequentibus scholis ... uocatur*, xv.2 *de coactili conficitur, de mollibus lanis ... composuit*, xvi.2 *propter aquam de puteis follibus hauriendam*, xviii.10 *habet de nauigii facilitate remedium*: for *a* in xxi.1 *de tua serenitate remedium*: and for one or other of these at v.7 *de summa integritatis intercipit*, where the *de* seems to clash hideously with the compounding element of the verb. The most daring use of the preposition comes at xv.3 *de libycis ... pellibus ... faciem*, where the prepositional phrase forms an analytic genitive case. The mode of expression may well have been influenced by the persistence in the writer's mind of *de coactili* and *de mollibus lanis* from the previous sentence, but the result inevitably recalls such analytic colloquialisms as Egeria's *de heramento et de argento modica frustella* (*itinerarium* 14.2).

c] e, ex

A surprisingly unorthodox, not to say incorrect, use of *ex* appears in ii.3 *ex auro ... erogandi diligentia*, an over-compressed expression. *ex auro* and *erogandi* must cohere, though at first sight of the immediate context the reader might legitimately think that we are dealing with an adnominal *ex* with causal force, linking *caecitas* with *aurum*, 'a blindness caused by gold', so completely does the syntactical articulation override the lightly-marked rhythmic pattern before the parenthesis: *a caecitate quadam ex auro, quod pretiosius habetur, profusior erogandi diligentia fuit*. I have not seen a parallel to the precise meaning of *ex auro erogandi* here, because *ex*, when used with *erogare*, would naturally denote the source, not the material, of the payment: so in the common classical formula *erogare ex aerario*. Conceivably *ex auro* was suggested by the compound *e-rogare*; but one cannot be sure. The author has in any case all but lost control of this sentence.

d] in

It is enough to note that *fieri in* (x.2 *fit autem ex ligno in modum sagittae facto*, xvi.2 *utres ... fiunt in magnitudinem trium et semis pedum*; both idiosyncratic usages, with the verb in a slightly different sense in each) is a collocation of overwhelmingly late attestation. As for case-usage after the preposition, at one point in the text (xii.2 *cultris in ordinem exstantibus*) *in* takes an accusative where an ablative would have been predicted; and the preceding *in pugnam* may have been responsible for this.

e] per

is ambiguous at vii.3 *habet foramina per quattuor partes*: does it apply to *habet* or to *foramina*? If the former, it acts distributively; if the latter, it stands for *in* or *trans*. Elsewhere in the text the preposition is fairly evenly divided between spatial (xvi.3) and instrumental (xiv.5) usages.

# II VOCABULARY AND COMPOSITION

## 1 LEXICAL ITEMS: SENSES AND SPELLINGS

*Words found only in the* anonymus *are marked with an asterisk*

### *ascogefyrus

(xvi picture-caption, chapter-heading; xix.3). On the whole a more successful compound (ἀσκός + γέφυρα) than the mongrel *currodrepanus* or the clumsy *thoracomachus*. The spelling with -f- is certain; see especially my critical note to xix.3. Greek is the natural language of ancient scientific terminology, and hence, presumably, the *anonymus'* predilection for his grotesque but endearing formations.

### *clipeocentrus

(viii-ix picture-caption, ix chapter-heading). A compound of mixed ancestry, *clipeus* + κέντρον (for κέντρον in the sense 'rivet' see Pausanias 10.16.1). The masculine inflexion is perhaps due to the continuing influence of the *clipeus* element, as *currus* may persist into the gender of *currodrepanus*; though this explanation fails in the cases of *ascogefyrus* and *thoracomachus*.

### cliuanus

So spelt by Σ at xv.2, with spirant for stop. The weight of contemporaneous evidence is overwhelmingly on the side of the etymologically-approved (κλίβανος) -b- spelling (Ammianus 16.10.8 *clibanarios* codd., 16.12.22 *clibanario* codd., historia augusta *Alex. Seu.* 56.5 *clibanarios* codd., codex theodosianus 14.17.9.3 *clibanariorum* cod.), though two manuscripts of Vegetius write *cliuanarii*(*s*) in 3.24 (p.117.5 Lang), and Isidore's etymology, *cliuanus a cliuo dictus* (*orig.* 15.6.5), points to an invasion by the -u- form. The *spirensis*, however, was absolutely consistent in spelling derivatives of *clibanus* with -b- wherever they appear in the *notitia* (or.v.29, v.40, vi.32, vi.40, vii.31, vii.32, vii.34, xi.8, xi.22, xi.26, xi.28; oc.vi.24, vi.67, vii.185, ix.33) and is unlikely to have itself introduced the phonetic corruption in a single occurrence of the basic noun in another text. *cliuanus* in *d.r.b.* xv.2 should therefore go back beyond Σ; perhaps only to Σ's ancestor, but perhaps further, and, *nolente Scriuerio*, I allow the transmitted spelling to stand as possibly authorial. The word seems to have been assimilated into *sermo castrensis* by way of the Greek-speaking East: Ammianus 16.10.8 *catafracti equites, quos clibanarios dictitant* (sc. Persae); historia augusta *Alex. Seu.* 56.5 (*'ex actis senatus'*, claims the writer, but the resemblance to the phrase of Ammianus just quoted is suspicious) *catafractarios, quos illi* (sc. Persae) *clibanarios uocant*.

### cunctus

Delicate shades of the concept 'all' are dissolved in this adjective, which replaces *totus* at pr. 15 *cunctam rem publicam* and *omnis* at ii.2 *cunctorum...cupiditates*.

### *currodrepanus

The components are occasionally divorced and separately inflected in the manuscripts (discussion in my critical notes to c. xii picture-caption and chapter-heading, xiv picture-caption), but it is practically certain that we have here a genuine though barbarous compound of *currus* and δρέπανον (or δρεπάνη), formed perhaps by tenuous analogy with the well-established Greek expression ἅρμα δρεπανηφόρον (Xenophon, *anabasis* 1.7.10, 1.8.10; Polybios 5. 53.10; Diodoros 17.53 *sub init.*; Lucian, *Zeuxis* 8 *med.*; and compare the Souda entry, δρεπανηφόρα τέθριππα). Fulgentius (*uirgiliana continentia* 752, p.94.14-15 Helm) supplies a fantastic etymology for *drepan(u)s* as a proper name: *Drepanos enim quasi drimipedos, drimos enim acer dicitur, pes uero puer uocatur.*

### *delectabilitas

In a sense in which the root meaning is strongly felt, at i.2 *opum delectabilitas*, the 'wideness of choice' of wealth, i.e. the 'profusion' of wealth. Schrijver proposed to replace the word by *delectabilitas*, which, in addition to the dubious recommendation of being itself ἅπαξ, makes nonsense of the text in the immediate presence of *nullo uetustatis more*.

### desperatus

Without verbal force, 'desperate' not 'despaired-of', in xviii.8 *contra desperati hostis audaciam*. A very rare sense in early prose (Cicero, *in Catilinam* 2.5.10 *desperatorum hominum flagitiosi greges*) becoming slightly more common later (Augustine, *in psalmos* 67.31 *desperatissimi peccatores*). Isidore (*origines* 10.72) indicates for his age a widespread acceptance, perhaps a colloquial quality, of the developed sense: *desperatus uulgo uocatur malus ac perditus nec iam ullius prosperae spei.*

### famosus

Probably in the common later sense 'famous', not 'infamous', in xviii.5 *Danubii, famosi pro latitudine fluminis* (compare e.g. Ammianus 23.6.65 *duo famosi nominis flumina, Oechartis et Bautis*: Martianus Capella 6.641 *Formiae etiam Laestrygonum habitatione famosae*. Petronius, at 70.10 *prasinianus...famosus*, neatly shows the incipient ambivalence of the word). Or so I take it in translating *d.r.b.* xviii.5; but some relic of the bad sense may still survive in the word here, and it could be translated 'notorious' without appreciable damage to the thought. No shade of the pejorative sense is present in the *famosa regna* of ii.8. Whatever the meaning of the adjective at xviii.5, its construction there, *famosus pro*, is confined to the *anonymus*: it is usually constructed, in all senses, with a plain ablative of cause or respect (Pliny, *epist.* 2.11.1 *personae claritate famosum*); and in ii.8, *antiquae paupertatis famosa regna*, although *famosus c. gen.* is once found (Appuleius, *met.*10.11 *mandragoram...grauedinis compertae famosum*), *antiquae paupertatis* is descriptive genitive with the noun.

### figurare

(i.5,6,8; *figuratio* iii.4; but *figura* in a different though allied sense in i.8 and iii.1). As applied to the striking of coins, a distinctively late usage (codex theodosianus 9.22.1.7 *figuratum solidum*, 12.7.1.3-4 *solidos ... nostris uultibus figuratos*). The *anonymus'* construction in i.5 *aeris quoque copiam in simulacris propriis figurabant* is unique: a plain ablative is normally used with the verb (historia augusta *Alex. Seu.* 25.9 *Alexandri habitu nummos plurimos figurauit*), and a construction with *in* is far less strained at codex theodosianus 9.21.7.2-3 *quidquid ex auro hominum priuatorum in monetis publicis reppereris figuratum* ...

### *fixorium

(viii.2; ix.1). More probably neuter than masculine with *clauus* understood. A special type of nail, perhaps a rivet, for which *clauus* would be inadequate, must be intended: the Paulus *ex Festo* gloss (p.208a Lindsay) *fixulas: fibulas*, if secure, provides the faintest of parallels to form and meaning.

### *fulminalis

(xviii picture-caption, chapter-heading, 6, 11). The form is found here only; *fulmineus* is usual. The *anonymus* is not alone in likening the blinding suddenness and explosive power of his machine's operation to a thunderbolt: Vegetius (4.22, p.144.2-3 Lang; the material re-worked by an interpolator in c.29, p.149.1-2 Lang) writes of the *onager* that *quanto amplior fuerit, tanto maiora saxa fulminis more contorquet.*

### liburna

(pr.12; xvii picture-caption, chapter-heading,1,3; xviii.9). 'Acro' comments, on Horace *epod.*1.1 *ibis liburnis*, that *liburnae nauiculae sunt textae uiminibus*, and the Paris scholia λ φ ψ *ibid.* describe their function in a sea-battle: ... *cum liburnis, id est nauiculis libycis quibus utuntur negotiatores Libyae: et hi etiam qui proelium nauale ordinant cum his nauiculis discurrunt inter maiores naues, ordinantes milites per singulas naues.* Vegetius gives more facts and supplies the etymology of the word: *diuersae autem prouinciae quibusdam temporibus mari plurimum potuerunt, et ideo diuersa genera nauium fuerunt. sed Augusto dimicante actiaco proelio, cum Liburnorum auxiliis praecipue uictus fuisset Antonius, experimento tanti certaminis patuit Liburnorum naues ceteris aptiores. ergo similitudine et nomine usurpato ad earundem instar classem romani principes texuerunt: Liburnia namque Dalmatiae pars est Iadertinae subiacens ciuitati, cuius exemplo nunc naues bellicae fabricantur et appellantur liburnae* (4.33, p.151.11 *sqq.* Lang: perhaps interpolated at the end). If *liburna* always connoted a small, lightly-built, fast galley, the word is conspicuously ill-applied to the ponderous ox-powered paddle-ship of *d.r.b.* xvii.

## plumbata

(*pl. tribolata* or *tribulata*, x-xi picture-caption, x chapter-heading, xi.1; *pl. mamillata*, x-xi picture-caption, xi chapter-heading). Plural instances in Vegetius (1.17, p.19.21 - p.20.1 Lang: *plumbatarum quoque exercitatio, quos mattiobarbulos uocant, est tradenda iunioribus*) and codex theodosianus (9.35.2 §1, where the *plumbata* is seen as an instrument of torture: *plumbatarum uero ictus, quos in ingenuis corporibus non probamus, non ab omni ordine submouemus, sed decemprimos tantum ordinis curiales ab immanitate huius modi uerberum segregamus, ita ut in ceteris animaduersionis istius habeatur moderatio commonentis*. Compare 11.7.3.1-3: *nemo carcerem plumbatarumque uerbera aut pondera aliaque ab insolentia iudicum repperta supplicia in debitorum solutionibus uel a peruersis uel ab iratis iudicibus expauescat*). Of the two epithets applied to *plumbata* by the *anonymus*, *mamillata* is disarmingly frank, and the participial-adjectival formation is unique (*mammatus* or *mammeatus* would be more regular [*mamillanae* with *ficus* in Pliny *h.n.* 15.19.69 means 'breast-*shaped*'], and *mamillare* [verb] is not recorded). *tribolata* or *tribulata*, however - on the spelling-variants and a textual problem see my critical note to the picture-captions and heading of c. x - *may* be found elsewhere if the older editors are right in following the ς reading at Palladius 1.42.3 *falciculas breuissimas tribulatas* (so ς: *tubulatas* DPKLSM *tabulatas* JG): but that reading can only be sustained if we attribute to it a sense ('sharp', 'pointed', even 'jagged') adhering far less literally to the base *tribulum* than that employed by the *anonymus*.

## quadrirotis

(vii picture-caption, chapter-heading). A recently-restored entry in Diocletian's *edictum de pretiis* (cap. 15.42a Giacchero = 15.38a [Latin corrupt] Lauffer: 63 *Journal of Roman Studies* (1973).99-110 at p.102) gives the Latin word, *currum quadrirotem*, not recovered by the editors from the already-known parallel entry in the Greek, κάρρον τετράτροχον.

## sequentior

With *aetas* in i.8, pushing to its limit the natural expression *sequens aetas* (at, e.g., Pliny *h.n.* 23.1.20). There seems to be no other recorded instance of the comparative of *sequens*.

## *thoracomachus

(xv picture-caption, chapter-heading, 1, 3 *bis*, 4; xix.2). Ineptly formed from θώραξ (in which sense? the breast or the breastplate?) and the root of μάχομαι. Saumaise (cited in Thompson's appendix) proposed to read *expositio thoraconacti* and would presumably have extended the correction throughout: it is attractive enough (τὰ νακτά [Hesychios] = 'felt': *de coactili...conficitur, d.r.b.*xv.2), but the number of occasions on which the change would have to be made is against it - unless we suppose a rigorous process of interpolation of the *-mach-* form at all points - and I suspect that it credits the *anonymus*, quite unjustifiably on his observed performance, with a knowledge of Greek equal to Saumaise's own.

*tichodifrus

(viii-ix picture-caption; viii chapter-heading, 1, 2; ix.1; xix.5, 6). The effect of compounding the dry and prosaic τεῖχος with the poetically-coloured δίφρος (admittedly found in a military context at Arrian, *taktika* 19.3) is bizarre. The spelling is certain, and the gender of the second Greek element probably explains the Latin masculine.

uis

carries an uncommon sense in *aurum argentumque et lapidum pretiosorum magna uis* (ii.2), but there is a close similarity here with Cicero's *magna uis auri argentique* in *tusc. disp.* 5.32.91. The resemblance between the two phrases is probably accidental; even less does xix.8 *longus machinarum siue tormentorum ordo* look like a conscious reminiscence of the Vergilian tag *ordine longo* (at e.g. *aeneid* 11.143-4), and the direct quotation from 'an excellent orator' reported in praef.5 remains unidentified and - despite Roman oratory's *penchant* for stating, in lapidary language, the blindingly obvious - may be fictitious.

uotiuus

in the striking phrase *uotiua direptio* at iv.4 *fin.*, takes the passively extended sense 'desired, longed-for, eagerly-awaited' earliest recorded for Appuleius (e.g. *met.* 7.13 *init.*) and Tertullian (e.g. *apologeticum* 9.2, ...*Tiberii, qui ipsos sacerdotes in eisdem arboribus templi sui obumbratricibus scelerum uotiuis crucibus uiuos exposuit*).

## 2 ABSTRACTION AND SUBSTANTIVAL STYLE

### a] abstract nouns

The *anonymus* thinks in abstractions: abstractions of time (*antiquitas, posteritas, uetustas*), abstractions of extent (*capacitas, densitas, enormitas, extremitas, latitudo, magnitudo, medietas, mediocritas, populositas*), abstractions of ethical values (*diuinitas, honestas, improbitas, iniquitas, libertas, pietas, prauitas, ueritas*), abstractions of inherent nature (*aequalitas, auxilium, diuersitas, ingenium, integritas, qualitas*) and of attributed quality (*admiratio, asperitas, caecitas, calliditas, celeritas, clementia, commoditas, cupiditas, difficultas, dignitas, facultas, felicitas, firmitas, fragilitas, frugalitas, imbecillitas, largitas, liberalitas, licentia, maiestas, necessitas, nobilitas, nouitas, paupertas, potestas, prouidentia, rotunditas, sagacitas, simplicitas, soliditas, sollicitudo, tarditas, temeritas, uilitas, utilitas*). Even from this incomplete list it can be seen that the abstracts in *-tas* easily preponderate over the rest; the text of about 3340 words contains no less than 54 different *-tas* abstracts (the count excludes *ciuitas* for obvious reasons) in a total of 118 inflexions. The favourite is *utilitas*, whose inflexions appear on fifteen occasions and in at least six different senses. (There are, besides, four instances of *utilis*). One word, *delectabilitas*, was coined to gratify this depraved taste, although elsewhere the *anonymus* seems to have missed his opportunities; he chooses to write *ad uiciniam pugnaturi hostis* in xix.6, where *ad uicinitatem* should have been equally possible, and *superior uita* is a forced and unconvincing substitute for *antiquitas* in i.4. Compelled to gloss his quasi-technical terms in c. xvii, he falls back on *-tas* forms: xvii.2 *in cuius alueo uel capacitate ... quarum supra ambitum uel rotunditatem ...*, where *rotunditatem* is recollected from xi.1 and *capacitas* appears for the only time in the text. The preceding sentence, xvii.1, contains *imbecillitas*, *utilitas* and *facilitatem* within the space of eight words.

The *anonymus* detaches himself from personalities, submerging them in abstraction: *dolosa barbaries* stands for *dolosi barbari* in vi.1, and the *possessorum distributa sollicitudo* guards the defence-walls proposed in c. xx. The abstract qualities of machines also take precedence over the machines themselves: *machinae huius admiratio* expresses 'this amazing machine' at xiv.1, and in xix.3 *ascogefyri auxilium* means not 'the assistance of the inflatable bridge' but the inflatable bridge itself, considered as a useful object. One may feel a slight tension here between the severely practical purposes of the devices and the abstract attributes which are chosen to represent and recommend them. But the uncertainty of the author's touch is revealed by the substitution of a more personal for an abstract quality in *uirtus* (not *uis*) *animalium* at xvii.1 *fin*.

### b] substantives and syntax

Reliance on substantives to carry the weight of the meaning has important syntactical consequences. So in the *anonymus*' phrase *quo protectior eadem ballista operetur* (viii.1) the dominance of noun over verb attracts the qualification from adverbial to adjectival form. Elsewhere we find an abstract noun used in preference to a verb-form, the subject of the avoided verb being given as a dependent genitive or adjectivally; *ballistae incessum* expresses *ballistam incedentem* in viii.1, and in xix.3 *militarem uiam* replaces *milites uiantes*. A more suspicious tendency appears in the clauses at ii.3 and vi.5 *profusior erogandi diligentia fuit* and *ut sit facilis imitandi confectio*: the verbs here are neutral, mere auxiliaries to the nouns on which the whole force of the expression is thrown, in a manner characteristic of the analytical tendencies of contemporaneous colloquial syntax, where, in its most developed form, strong nouns, weak verbs and prepositional indications of agency replace the synthetic dominant verb. Egeria (*itinerarium* 1.2) writes *consuetudo est ut fiat hic oratio ab his qui ueniunt* for *consuetum est ut hic orent qui ueniunt*: the *anonymus* is nowhere so informal or so revealing in his syntax as this, and yet his *profusior erogandi diligentia fuit* is a paraphrase of (it might be) *diligenter profusiusque erogatum est* (I am not responsible for the choice of tense here), and *ut sit facilis imitandi confectio* is an analysis of *ut facile conficiantur* (sc. *machinae*) *imitando*. And I have already called attention (above, section I.6.[b]) to an expression, *de libycis...pellibus...faciem* (xv.3), in which the *anonymus* analyzes a genitive case.

## 3 IMAGE AND METAPHOR

It is perhaps neither accidental nor inappropriate that the *anonymus* frequently employs, in a text dealing with reform and invention, the vocabulary of **birth** and **propagation** as a source of metaphor: so *ingenii...magnitudo, quae uirtutum omnium mater est* (praef.6), *nominis tui gloriam propagare* (ii.7; *qui romano nomini debitos affectus propagatis in filios* is barely metaphorical in praef.8), *ne profusa largitio semina magis excitet proeliorum* (i.1; the image in the verb is used with a strangely different emphasis in vi.5, *in excitandis armorum generibus*), *non amplius bellorum florebit improbitas* (i.1), *expensis grauidam rem publicam* (v.3; there may be some lexical confusion with v.3 *grauet* here), *ut...utres...non extollant uterum* (xvi.2), *impetu parturiente discursum* (xvii.2). **Light** and **darkness** are equally obvious and useful sources of imagery in such a context: *ne commixtionis licentia...integritatem publicae utilitatis obfuscet* (iii.2), *legum...sententias...iudicio augustae dignationis illumines* (xxi.1), and an extreme transfer of *illustrare* in the preface, 7 and 10, *cum barbarae gentes neque facundia polleant aut dignitatibus illustrentur* and (more curiously) *limitum solitudinem ...incola securus illustret*. Other metaphors are drawn from physical sensation: from **sound** in *pictura praenuntiante* (iii.4), *circumlatrantium...nationum...insania* (vi.1), *iustitia profitente* (xxi.1); from **bodily affliction** in *tributariae functionis laborat illatio* (v.1), *quietem pacis lacerant* (vi.3), *thoracomachus pluuiis uerberatus* (xv.3); from **weight** in *ne...fides...uelut fallacem in posterum grauet* (praef.3), *cumulatus honoribus miles exultet* (praef.10; the verb is not intended literally); and from **strength** in *ut diuina consilia diuinis successibus conualescant* (praef.1), *armorum uigorem* (praef.15), *res publica moribus ditata firmabitur* (iv.5; the image of 'enrichment' gives direction to the image of 'strengthening'). ii.2 *sqq.*, especially at 5 and 6, is a passage dominated by the image of **fire**; *...cunctorum dandi habendique cupiditates accendit... sed afflicta paupertas in uarios scelerum conatus accensa... inflammando odia... produxit <inopia> magis quam succendit audacia*. This follows a miscalculated **liquid** image, ii.1 *huius auaritiae origo hinc* *creditur emanasse*. An easy metaphor from an item of clothing is hesitantly submitted in xx.1, *ut prouinciarum quies circumdata quodam praesidii cingulo inlaesa requiescat*.

None of this is very new. Even *quies...circumdata...praesidii cingulo* is not so daring as the author thinks; Appuleius approaches it in *de mundo* 7.304, *Asia ... constringitur ... oceani cingulo*. And the rest are obvious enough. Amongst his more forceful metaphors the *anonymus* may be considered to have been most successful with the *circumlatrantes nationes* who *quietem pacis lacerant*, images of a violence appropriate to the context at vi.1-3; with the first, compare the *circumlatrantes philosophi* of Lactantius *diu. inst.* 2.8.50 and the *circumlatrantes haeretici* of Augustine *epist*.65.1, and with the second, Seneca's phrase *ad...pacem...lacerandam* in *dial*.4.35.5. But elsewhere one may wish to question the extent to which the author was aware of the metaphorical quality of some of the words which he chooses to employ. Every current vocabulary contains subdued, suppressed or even dead metaphors, and Livy 1. 58.7 *uestigia uiri alieni, Collatine, in lecto sunt tuo* stands as a frightful warning of what can happen when a latent metaphor is ignored. In the case of at least one word the *anonymus* displays that coarseness of diction which results from the over-riding of a concealed image (and also that which results from an imprecision of meaning). Strictly speaking, the scythes are not 'unfolded' from the stub-axles of the *currodrepanus*, nor is the *ascogefyrus* actually 'unfolded' from bank to bank of a river, and yet *explicare*, the *mot juste* for neither operation, is used for both in xiv.4 and 5 *licet conuertere...explicatas falces... funis uero, quo laxato...explicantur falces* and in xvi.3 *hoc idem opus...ad alteram facile explicabitur ripam*.

What is also noticeable in the *anonymus'* style is the near-personification of abstracts achieved by coupling them as subjects to personal verbs. From such natural expressions as the *slowness of promotion* which *discourages* men from army service (v.5) and the *matériel* which *provokes* an engagement (xix.1), we pass to an abstracted *afflicted poverty* which *has before its eyes* no respect for law or honour (ii.5), an *ingenious necessity* which *discovers* (xvi.1), a *utility* which *summons* (xvii.1), the *operation* of machines which *repels* (xix.4), a *rotundity* which *has power* to penetrate obstacles (xi.1), a *yielding weight* of the enemy which *meditates* flight (xix.7), and at last we find explicit personification; the ballista, *guardian* of the frontiers (xviii.11), the *fearful apprehension* which, with *cleverness* as its *mistress*, *contrives* the woollen shirt (xv.2), and the *justice* which *delivers its ruling* (xxi.1). It is a curious consequence of the late retreat into abstraction that the possibilities of imagery should be so strikingly extended.

## 4 PROSE RHYTHM

The *anonymus* employs five accentual clausulae, the three normal basic types and a standard variant of each of two of these. The patterns are:

(1) ´°°´°, where ´ denotes an accented and ° a slack syllable. I will call this the '(cursus) planus 1', abbreviated to p1 in the analysis which follows.

(2) ´°°°´°, 'planus 2' or p2, a common extension of planus 1. The third syllable very often carries a secondary accent, ´°`°´°.

(3) ´°°´°°, 'tardus 1' or t1.

(4) ´°°°´°°, 'tardus 2' or t2, an extension of tardus 1. The third syllable may fall at a secondary word-accent, ´°`°´°°.

(5) ´°°°°´°, 'velox' or v. The fourth syllable almost always carries a secondary accent, ´°°`°´°.

The incidence of these patterns at the major pauses in the rhetorical articulation of the text is as follows:

praefatio 1: *uestrae commoditas* t1 *-tissimi principes* t1 *-renda temporibus* t1 *-cessibus conualescant* v.

praefatio 2: *-bello composui* t1 *-tati sufficiat* t1 *-mento praemisso* p1 *fides ostendi* p1.

praefatio 3: *posterum grauet* p1 *praemii loco* p1 *cesserit, posco* p1 (*<praemium> prosequatur* v) *-dacia declinare* v.

praefatio 4: *-rata cognoscere* t1 *lateat inquirentem* v.

praefatio 5: *fuerint approbati* v *optimus orator* p2 *-tura donauerit* t1.

praefatio 6: *sentiat quam loquatur* v *litteris acquisita* v *-netur inuentio* t1 *omnium mater est* t1 *-tate subnixa* p1.

praefatio 7: *-demus accidere* t1 *-tatibus illustrentur* v *-bentur alieni* p2.

praefatio 8: *-tate diligitis* t1 *-gatis in filios* t1 *-tatis intulerit* t1.

praefatio 9: *uestrae tractantibus* t1 *-uato contentis* p1 *terrae cultoribus* t1 *lucra tractantibus* t1 *-tate gaudentibus* t1 *commoda singulorum* v *-iecta testabitur* t1.

praefatio 10: *cultor habeatur* p2 *-sante contumelia* t2 *-curus illustret* p1 *poena duplicetur* p2 *miles exultet* p1.

praefatio 11: *-toriis procurentur* v *-uenta referemus* p2.

praefatio 12: *-terio praeualere* v *turbae obruantur* p2.

praefatio 13: -*tata sollertia* t1 -*turus inuadat* p1 *arte muniatur* p2 *hostium strages* p1.

praefatio 14: -*oni onerosa* p2 -*busque necessarium* t2 -*menta portabunt* p1.

praefatio 15: *arbitror designatis* v -*tatis apporto* p1 -*mediis subleuandam* v *uestrae clementiae* t1 *nobis aliena* p2.

praefatio 16: -*patos effugiunt* t1 -*tatibus peregrinus* v -*ferre gestiui* p1.

praefatio 17: -*tate protulerit* t1 *uenia protegendum* v -*sophiae libertatem* v.

i.1: (-*atur aerarii* t1) *excitet proeliorum* v -*tatis reprimatur* p2 -*rebit improbitas* t1 -*sidia recreantur* v.

i.2: -*uanda profuderit* t1 -*tatibus subuenire* v.

i.3: -*dentia referenda* v -*genis habuerint* t1 -*centia uideatur* v.

i.4: *moenibus laetabantur* v -*teriam conferebant* v.

i.5: -*monia figurabant* v.

i.6: -*tati suppeteret* t1 *igne solidatam* p2 -*onibus figuratam* v *usui habuerunt* v.

i.7: -*tatis inuentum* p1 *modico signauerunt* v -*toribus utebatur* v.

i.8: -*dante materia* t1 -*sabat ornatus* p1 -*menta meditata* p2 -*osius figurauit* v -*turnior fuit* p1 -*toque signauit* p1 *usibus proficiens* t2 -*crata permansit* p1.

i.9: -*mercia signabatur* v.

i.10: -*cilius prosequatur* v -*poribus diuersas* p2 -*tatibus properaui* v.

ii.1 *pretii habebatur* v -*merciis assignauit* v *creditur emanasse* v.

ii.2: *templis reposita* t1 *publicum peruenisset* v -*tates accendit* p1.

ii.3: *uultu signatum* p1 -*gatio uideretur* v -*osius habetur* v -*gentia fuit* p1.

ii.4: -*ores effectae* p1 -*lentia oppressis* p2.

ii.5: -*natus accensa* p1 *oculis habens* p1 *artibus commendauit* v.

ii.6: -*fecit imperia* t1 -*ciniis persequendo* v *fouit tyrannos* p1 -*cendit audacia* t1.

ii.7: -*dentiae tuae* p1 *optime imperator* v -*tori prospicere* t1 *gloriam propagare* v.

ii.8: -*uolue memoriam* t1 *regna considera* t1 *opibus norant* p1 -*rupta commendet* p1.

ii.9: *penitus non habebant* v.

iii.1: *publicae non ferenda* v *fraudibus deprauata* v -*one sollicitat* t1 -*tatis imaginem* t1 -*tatur, imminuit* t1 -*lenta calliditas* t1 -*nosa necessitas* t1 -*tractibus intulerunt* v -*esse simplicitas* t1.

iii.2: *-benda correctio* t1 *insulam congregentur* v *usibus profuturi* v *terrae prohibiti* t1 *fraudibus opportuna* v *-tatis obfuscet* p1.

iii.3: *-netae praestabitur* t1 *mercis occasio* t1.

iii.4: *-onis appareat* t1 *-ante subieci* p1.

iv.1: *artibus uexant* p1 *-cranda cupiditas* t1 *-tatibus inimica* t2 *-rentia dignitatum* v *missos existimant* t1 *-cedit iniquitas* t1 *-rari medicina* p2.

iv.2: *possit iniquitas* t1 *uires exhauriant* t1 *soli peccarent* p1.

iv.3: *-lorum illibata est* t2 *praeda discessit* p1.

iv.4: *-ronum comparatio* t2 *-menti coemptio* t1 *moenibus profutura* v *-tiua direptio* t1.

iv.5: *cupidi moderentur* v *-tata firmabitur* t1.

v.1: *publicae incommodis* t2 *merito repellendis* v *-benda ueniamus* p2 *-borat illatio* t1.

v.2: *-gusta fastidiat* t1 *exitum declarabo* v.

v.3: *aliquot peractis* p2 *-menta peruenerit* t1 *publicam grauet* p1 *gaudeat absolutus* v *-terior succedens* p2 *-uissimis releuabit* v.

v.4: *locum uocatur* p1 *-natus abscedat* p1 *-turus accedat* p1.

v.5: *publicam recreabit* v *-onis imminuet* t1 *tarditas prohibebat* v.

v.6: *-onis utilitas* t1 *donis opulentos* p2 *-lentes agricolas* t1 *-fenderant loca* p1 *milite collatores* v.

v.7: *-lorum ruina* p1 *-serta militia* t1 *-tatis intercipit* t1 *damna supplenda sunt* t1 *-geni iuniores* p2 *-triculis continentur* v *armis exerciti* t1 *-pendio subleuati* v *tulerit subrogandi* v.

v.8: *ita prouisis* p1 *-nebit exercitus* t1 *primis subsidia* t1.

vi.1: *-stringat insania* t1 *-losa barbaries* t1.

vi.2: *siluis teguntur* p1 *-cantur pruinis* p1 *nimio proteguntur* v.

vi.3: *facile queunt* p1 *-atis incursibus* t1.

vi.4: *modi nationes* p2 *moenibus defenduntur* v *machinis prosequendae* v.

vi.5: *-neribus oriatur* v *uero distantem* p1 *-oni subieci* p1 *-tandi confectio* t1.

vii.1 *-emplum ballistae* p1 *-tura testatur* p1.

vii.2: *quattuor facilitas* t2 *bellicos trahit* p1 *artis industria* t1 *-gittas impellat* p1 *manus imitata* p2.

vii.3: *quattuor partes* p1 *-ducta et flexa* p1 *-rata consistat* p1.

vii.4: *-ponitur celerius* t2 *-rigitur subleuata* v.

vii.5: *-cessitas uocet* p1 *-flexus erigitur* t1.

vii.6: *opera uirorum* p2 *-torta iaculatur* p2.

viii.1: *sumpsit exordium* t1 *-retur ascensus* p1 *-cendus incessum* p1 *-lista operetur* p2 *-sitio uel fabrica* t2 *-one praeparanda* p2.

viii.2: *-milior fabricatur* v *-lare uestigia* t1 *-isque confixis* p1 *-one defendit* p1 *-tenter impulsus* p1.

viii.3: *-perior latitudo* v *-genter aptatis* p1 *-atur facultas* p1.

ix.1: *modicus clipeus* t1 *-genter munitus* p1 *-betur idoneus* t1 *operit gestientem* v *-difri oppositus* t1 *-mone suspensus* p1 *-ferius latitantes* v *comminus uenientum* v.

x.1: *iaculi genus* p1 *-detur instructum* p1 *-sueuit emitti* p1 *comminus uadit* p1 *-ferre perniciem* t1 *perimit infixum* p2 *-cantis infigitur* t1 *fuerit conuersum* p2 *inferat uulnus* p1.

x.2: *-tatum infigitur* t1 *-rumper extensa* p1 *tribuli emergunt* p2.

x.3: *-nentis amplecti* p1.

xi.1: *pennis adhaerentibus* t2 *-tate adiuta* p1 *ualeat penetrare* v.

xii.1: *-hiculi genus* p1 *-detur instructum* p1 *pugnae necessitas* t1.

xii.2: *-genter muniti* p1 *-stantibus communitur* v *-atur ascensus* p1.

xii.3: *currus aptantur* p1 *ansulas habentes* p2 *-xati quidem explicant* t2 *erigunt falces* p1.

xii.4: *hostibus immittant* p2 *strages efficiant* t1 *bella cognoscunt* p1.

xiii.1: *tractu discurrit* p1 *-tute protegitur* t1 *-renda pericula* t1 *-toribus gubernatur* v.

xiv.1: *parte dissimilis* t1 *ferro circumdatis* p2 *positis communitur* v.

xiv.2: *-one dissimilis* t1 *-gacitas regit* p1.

xiv.3: *-tate secernitur* t1 *minus amplectitur* t1.

xiv.4: *axe porrectas* p1.

xiv.5: *axes eriguntur* p2 *parte exteriore* v *-untur incursus* p1.

xv.1: *-tauit antiquitas* t1 *-tati subiecit* p1.

xv.2: *-mani conficitur* t1 *-gistra composuit* t1 *-tate non laederent* t1 *-misque discrimen* p1 *-bori sufficiant* t1.

xv.3: *pluuiis uerberatus* v *-ficiat uestientem* v *-usdem thoracomachi* t2 *superinducere* t1.

xv.4: *-comacho inducto* p2 *nomen assumpsit* p1 *ocreis inductis* p2 *lateri aptato* p2 *miles armabitur* t1.

xvi.1: *-tineri exercitus* t2 *-cursus impediat* t1 *-pendio facile* t1 *usu praecipuum* t1 *-osa necessitas* t1 *-one componitur* t1.

xvi.2: *arte mollitis* p1 *follibus hauriendam* v *-flati tumuerint* t1 *plana perficiat* t1 *inuicem colligantur* v *circulis exstantibus* t2 *-tuntur uncini* p1 *pontis adsociata* v *-catur integritas* t1.

xvi.3: *impetum meatus* p2 *-cabitur ripam* p1 *ferreis palis* p1 *desuper extensis* p2 *-undi per fluuium* t1 *-tineris apparatu* v *temporis spatium* t1 *liberam facultatem* v.

xvi.4: *-terea sumus* p1 *-nenda uestigiis* p1 *deneget firmitatem* v.

xvi.5: *-listae dispositae* t1 *-mento consistat* p1.

xvii.1: *-doneam bellis* p1 *-mana prohibebat* p2 *-tilitas uocet* p1 *ope subnixa* p1 *uirtus impellit* p1.

xvii.2: *machinis adiuncti* p2 *-teribus uoluunt* p1 *-natibus elidentes* v *-fectu operantur* p2 *-ente discursum* p1.

xvii.3: *semet operantibus* t2 *pugnam capescit* p1 *comminus uenientes* v *-tritu comminuat* t1.

xviii.1: *-oni necessarium* t2 *usu compertum est* t1 *-primitur erecto* p2 *-missus impellit* p1.

xviii.2: *militum trahi* p1 *-tudo permittit* p1 *nisibus adnitentes* v *machinis acquisitis* v.

xviii.3: *-milius tela* p1 *uocet utilitas* t1 *erigit nunc deponit* v.

xviii.4: *-tutis argumentum* p2 *-tate connexum* p1 *opera gubernat* p2 *-terio inseruiret* v *artis inuentio* t1.

xviii.5: *artibus communita* v *longius uadit* p1 *-tudine fluminis* t1 *ualeat penetrare* v.

xviii.6: *etiam nuncupata* v *-tatur effectum* p1.

xviii.7: *-uicte imperator* p2 *-cabis exercitus* t1 *uiribus obuiando* v *artis ingenio* t1 *-pereris ualituras* v.

xviii.8: *-unda discrimina* t1 *hostis audaciam* t1 *agmina persequeris* t2 *-culta temeritas* t1 *ante quam ueniat* t1 *-atus impediet* t1 *-curret inuentio* t1.

xviii.9: *hostis obsideat* t1 *-tatis ingenio* t1 *bubus subacta* p1 *-burna transcurrens* p1 *mora uictoriam* t1.

xviii.10: *uiribus resistet* p2 *nititur firmitate* v *-tate remedium* t1.

xviii.11: *limitum custos* p1 *-sterium quaerit* p1 *-tute praestantior* t1 *-ferre potuisset* p2 *-tata largitur* p1.

xix.1: *prouocet certamina* t2 *-pressa minuatur* p2 *-minibus ordinare* v *milibus instruantur* v *-pressa populositas* t2 *-xiliis abundans* p2 *copiis opprimatur* v.

xix.2: *militem plagas* p1 *-comacho muniri* p2 *-nentibus rebus* p1 *-tilitas armari* p2 *-goribus sufficiat* t2 *possit occurrere* t1 *clipeis assumptis* p2 *-rumque euenit* p1 *densitas inaccessa* v *-tate refugium* t1.

xix.3: *-tina subsidia* t1 *-tatum uehiculis* t1 *-tetur auxilium* t1 *-retur occursus* p1 *res ita tulerit* t1 *-cilitas offeratur* v.

xix.4: *ceteras nationes* v *suppetit uirtus* p1 *-ratu superanda* p2 *-dustria proponenda* v *uiribus resistat* p2 *gentis refutet* p1.

xix.5: *-sterio uirorum* p2 *-sidiis profuturam* v.

xix.6: *tegmine latentes* p2 *hostis aduenient* t1 *ferreis communitis* v.

xix.7: *hostium pondus* p1 *-sidia meditetur* v *poplites incidantur* v.

xix.8: *-cari permiserint* t1 *ordo circumdet* p1 *hostis infestet* p1.

xix.9: *-tilior cura* p1 *-torum ducatur* p1 *dubiis rebus* p1 *casibus subuenire* v.

xx.1: *limitum cura* p1 *latus imperii* t1 *-stella prospicient* t1 *turribus erigantur* v *sumptu constituat* t1 *-grariis exercendis* v *-sidii cingulo* t1 *-laesa requiescat* p2.

xxi.1: *-uina prouidentia* t2 *-tissime imperator* v *-sidiis comparatis* v *-tate remedium* t1 *-rarum medicinam* p2 *-asque sententias* t1 *-iecto litigio* t1 *-onis illumines* t1 *-sistit alienum* p2 *-ceri certandi* p1 *-titia profitente* v *merita singulorum* v.

This conspectus gives the final rhythm of every sentence, together with the subsidiary falls in a number of complex sentences which show an unusually high degree of concern for the internal rhythmic structure. p1 and t1, in roughly equal proportions, are the most common cadences, with v in third place. t2 is infrequent, and is more likely to be found at an internal pause than at sentence-end, where p1 and v are favoured. Many sentences fall, just after their beginning, on a subsidiary close of p1 type. The pattern t1 often occurs in grouped succession, but the fine sequence of t1 cadences in *praefatio* 9, reflecting the parallel syntax of the clauses, may be an accidental effect of the choice of words; the run of internal t1 falls in iii.1 almost certainly is. In contrast with *praefatio* 9, parallel clauses in xiii.1 are not arranged in parallel rhythm. The clausulae of the most ambitious passages - the preface and xviii. 7-11 - show about as much variety within and between sentences as the more straightforward parts of the exposition; the rhythmic resources inadequately reflect the change of tone, and the most skilfully-organized passage is to be identified neither in the preface nor in c. xviii but at xvi.1-3. (xviii.8, like xvi.1, shows an obsessive persistence of rhythm which may be connected with the persistence of words and forms characteristic - as we shall see - of the author's style). Several paragraphs open with a sentence containing frequent and varied internal falls (iii.1, iv.1, ix.1, xvii.1), but thereafter the handling of rhythm seems less controlled: the amorphous x.2 disappoints after the taut construction of x.1.

Clausula helps to decide text at the last word of ix.1 (where CM's *uenientium* is *lectio facilior* and produces an impermissible cadence) and forbids Reinach's *recreabuntur* - avoided by the *anonymus* precisely because of its rhythmic form - at i.1 *fin.* Study of the clausulae seems to have no practical application elsewhere. Much of what we see may be no more than the result of an instinctive feeling for rhythmic shape; but the preferences and arrangements just noticed are a sufficient indication that the *anonymus* was concerned in some measure at least to cast his prose into acceptable rhythmic forms.

# III STYLISTIC DEVICES AND CHARACTERISTICS

## 1 SOME RHETORICAL FIGURES

### a] anaphora

The figure conveys the impassioned appeal of praef.8 ...*principes, qui gloriam bonae opinionis perpetua felicitate diligitis, qui romano nomini debitos affectus propagatis in filios* ...

### b] asyndeton

expresses dramatically the agitation of ii.6 *populando agros, quietem latrociniis persequendo, inflammando odia,* where the order is powerfully chiastic, although the *anonymus,* either from excess or deficiency of art, has not arranged his clauses in crescendo.

### c] chiasmus

Immediately preceding the chiastic asyndeton just noticed we find *nullam reuerentiam iuris aut pietatis affectum* in ii.5. The structure is slightly shaky: *nullam* in the leading phrase throws the intended opposition visibly out of balance, and equilibrium is uneasily restored by understanding *ullum* ἀπὸ κοινοῦ in the second limb.

### d] figura ἀπὸ κοινοῦ

A weak example at xiii.1 *unius hominis arte regitur et uirtute protegitur,* where the genitive is common to both clauses. On constructions implying an understanding ἀπὸ κοινοῦ in ii.5 and (perhaps) xv.2 see respectively [c] above and [o] below.

### e] figura etymologica

xviii.2 *radiorum nisibus adnitentes*: an inexpensive ornament in an ineptly constructed sentence.

### f] hendiadys

A single idea may be intended in *manus impetu et uiribus* (x.1) and in *sagittarii libertatem et manus* (vii.2); but the second of these is more convincingly explained as a zeugma ([o] below).

### g] homoioteleuton

Such riming terminations as those in v.4 *comparatio,* ...*coemptio,* ...*direptio* could be accidental; but *arte regitur et uirtute protegitur* in xiii.1 looks deliberate.

### h] hyperbaton

Only an extreme example need be noted, xvi.3 *transeundi* ... [fourteen words intervene] ... *facultatem.*

i] litotes

An instance, whose degree of force is somewhat indeterminate, in v.1 *ratione non incongrua.*

j] oxymoron

The collocation *deflexus erigitur* in vii.5 might be dignified by this title; but it is more probably an accidental harshness.

k] paronomasia

A writer with a finer sense of the choice and placing of words would not have allowed the momentary suspicion of a pun in xxi.1 *ad ciuilium curarum medicinam.* In xii.3 and 4 *funes...funera* the apparent play on sound seems, in view of the distance between the two items, to be a special case of the persistence of words in the author's mind (on which see below, section 2). This may also be the true explanation of the echoes at xvi.2 *utres...non extollant uterum* and xvii.3 *liburnas comminus uenientes...comminuat,* although in both these cases the words are physically close enough to admit the possibility of a deliberate play on sound.

l] personification

is discussed with Imagery, above, section II.3.

m] pleonasm

*uera fides* in i.10 and *integra fides* in iii.3 come dangerously close to redundancy of expression, the former perhaps more so than the latter; and there is a slight superfluity in *remedium ad ciuilium curarum medicinam* at xxi.1 which would have been less noticeable if *medicinam* had been turned verbally.

n] variatio

A particularly self-conscious example of this amongst the conjunctions in the preface, 10 *quem ad modum... unde... quo pacto... quo argumento*; and the phraseology for parallel concepts is very deliberately varied in xvi.2 and 3, *et est priori quidem currui hac ratione dissimilis... a secundo uero curru hac diuersitate secernitur...*

o] zeugma

makes a startling appearance in xv.2 *ad mensuram et tutelam humani pectoris*; unless one prefers to regard this as a variety of construction ἀπὸ κοινοῦ, with *ad* implied in a different sense with the second noun. Almost equally crude is the clear zeugma in xviii.9 *liburna...rotis ac bubus subacta*: but *sagittarii libertatem et manus imitata* is less prominent in vii.2, a sentence in which also the word-order of the phrase *duobus subiunctis et armatis equis* converts a plain expression into an almost zeugmatic collocation. The double subject in the transmitted text of i.6 *ut uero emendi et uendendi utilitas et ut facultas regiae largitati suppeteret,* on which see my critical note, is to be eliminated by emendation.

## 2 REPETITION OF WORDS

This I take to be the most remarkable characteristic of the author's style. The recurrence of particular words (*utilitas*, for example) at short intervals, sometimes with a change of sense, is sufficiently obvious at a first reading of the text; and close examination of the prose reveals how pervasive the repetitions are, extending beyond single words to whole phrases and constructions, and often forming sequences of considerable complexity, where recurrences of one item are interwoven with or overlap those of another. Postponing general comment until the end of this section, I give now, in condensed quotations, a survey of the most conspicuous repetitions of word, form, root, idiom, phrase and construction found in the course of the text, printing in bold type the recurring items.

[a] praef.2-3:

*de largitionum* **utilitate** ... *non quod istud tam inmensae* **utilitati** *sufficiat sed ut* ... *in reliquis possit* **utilitatis fides** *ostendi. proinde ne* **promissioni fides** *accommodata* ... *poenam praemii loco, si* **promissio** *secus cesserit, posco* ...

Notice the change in meaning between the first two appearances of *utilitas* and the last, and between the two examples of *fides*.

[b] praef.6-15:

*aut eloquentia litteris acquisita* **consecuta est** *utilitates artium* ... *neque facundia polleant aut dignitatibus* **illustrentur** ... *quae nostris sensibus* **commoda** *prouidentia* **diuinitatis** *intulerit* ... *pro saeculi uestri felicitate* **gaudentibus**, **consequemini** *ex hoc opere* **commoda** *singulorum* ... *erectis castrorum munitionibus incola securus* **illustret** ... *cumulatus honoribus miles* **exultet** ... *ingenii* **magisterio** *praeualere* ... **talis** *est excogitata sollertia* **ut** *equus* ... **tali** *arte muniatur* **ut** ... *sine cuiusquam* **magisterio** *efficiat* ... *munus concessu* **diuinitatis** *apporto.*

[c] praef.13 *with* xii.4:

(praef.13) *ut* ... **efficiat** *magnas hostium* **strages**.

(xii.4) *quas uero turbatis ordinibus* **strages efficiant** ...

[d] i.1 *with* ii.1

(i.1) *ne* **profusa largitio** *semina magis excitet proeliorum.*

(ii.1) *Constantini temporibus* **profusa largitio** ... *assignauit.*

[e] i.4-10, an exceptionally intricate passage which I transcribe *in extenso*:

4 *in quorum decorem uniuersam* **auri argentique materiam** *confere-*
5 *bant.* **aeris quoque copiam** *in simulacris propriis ad uirtutis*
6 *suae testimonia* **figurabant. ut uero** *emendi et uendendi utilitas*
*** *et* **ut** *facultas* **regiae** *largitati suppeteret, in* **aeris usum**
*excultam politius terram, certis* **quoque** *expressionibus* **figuratam,**
7 **auro argentoque** *reposito* **usui** *habuerunt.* **sed** *posteriores fasti-*
*dientes priscae uilitatis inuentum formatis e coriis orbes* **auro**
*modico* **signauerunt**, *quibus regum munificentia et commutationum*
8 *necessitas intactis collatoribus* **utebatur. sequentior uero**
*aetas,* **aeris** *redundante* **materia** *quam publicus iam moenium recu-*
*sabat ornatus,* **diuturna ut** *prior monumenta meditata* **aes** *ualidum*
*ipso* **pondere** *pretiosius* **figurauit**; *cuius* **species pro**
**qualitate, ut** *diximus,* **ponderis diuturnior** *fuit,* **sed**
*potentia* **regalis pro** *licentia* **speciem** *suam tantum in* **auro**
**argentoque signauit, quae pro** *reuerentia* **figurae** *nullis*
9 **usibus** *proficiens* **ad** *honorem* **regium** *sacrata permansit.* **aeris**
*autem* **materia, quae** *iam* **pro copia** *uilior erat,* **ad** *dona*
10 *militaria et* **uaria** *populorum commercia* **signabatur. sed ut**
*uera fides dicta facilius* **prosequatur**, *formas et* **species**
*commodas atque* **pro** *temporibus diuersas* **uarias***que ueterum*
*prouisiones exprimi colorum* **qualitatibus** *properaui.*

This is very arresting. Some of the repetitions were, perhaps, unavoidable: thus *aeris* in 5 and 6, *auro* in 6 and 7, *aeris* and *aes* in 8, since the *anonymus* cannot avoid mentioning gold and bronze at these points: and it is probably rather unfair to draw attention to the obvious formula *aurum argentumque* in its various inflexions at 4, 6 and 8. Some of the echoes may have arisen accidentally (so perhaps the two modal *uts* in 8, and the sequence *regiae - regalis - regium* [6-8-9]); some of the repetitions are quite deliberate (such as the two final *uts* in 6 and the calculated *uariatio* of *uero - sed - uero - sed* between 6 and 8). But such recurrences as *figurabant - figuratam - figurauit - figurae, signauerunt - signauit - signabantur, pro - pro - pro - pro - pro* and *diuturna - diuturnior* are indicative either of an obsessive habit of thought or of a chronic poverty of verbal invention. Observe also the preoccupation with comparative forms (mostly with genuinely comparative force) throughout the whole of the chapter, *superioris - posteriores - sequentior - prior - pretiosius - diuturnior - uilior - facilius*. Some items in the sentences printed above appear in the surrounding context: *collatoribus* in 7 recalls the inescapable *collatorum* of 1, and *licentia* in 8 picks up the same word in a different sense at the end of 3: one word, *largitio,* outside the main area of repetitions given above, appears twice earlier in the chapter (1 and 2). Sentence 10 terminates chapter i, and - significantly - with the pause in the train of thought the author breaks out of his obsession with causal *pro*: the word is still in his head, but in the new paragraph (ii.1 *aurum pro aere*) he can use it in another meaning.

[f] ii.2-6:

*cunctorum dandi habendique cupiditates* **accendit**. ... *enormis iam et grauis* **erogatio** *uideretur* ... *profusior* **erogandi** *diligentia fuit.* ... *sed afflicta* **paupertas**, *in uarios scelerum conatus* **accensa** ... *quos ad* **gloriam uirtutis tuae** *produxit* <*inopia*> *magis quam* **succendit** *audacia* ... *in posterum* **nominis tui gloriam** *propagare* ... *antiquae* **paupertatis** *famosa regna considera* ...

[g] iii.1-2:

*solidorum figura aliquantorum* **fraudibus** *deprauata* ... *ementis enim eundem solidum* **fraudulenta** *calliditas* ... *ne commixtionum licentia* **fraudibus** *opportuna integritatem publicae utilitatis obfuscet* ...

[h] iv.1-2:

*quae prouincias auaritiae* **artibus** *uexant, accedit etiam iudicum execranda* **cupiditas** ... *ab his procedit* **iniquitas** *unde debuit sperari medicina. et tamquam sua* **rebus** *sufficere non possit* **iniquitas**, *exactores in profligandis* **rebus** *huius modi dirigit unus quisque qui diuersis rapinarum* **artibus** *collatorum uires exhauriant.* ... *quod si prouincias casti et integritatis* **cupidi** *moderentur* ...

Here I might also have drawn attention to *auaritiae* and *rapinarum* before the two appearances of *artibus*. The repetition of *rebus*, which I take to be the essential reason for deleting the word between *sua* and *sufficere*, is in fact an argument for retaining it in the text. Similarly the presence of *otio* in v.3 makes the restoration of *otio* for †*talio*† at v.4 more, not less, attractive.

[i] v.3-5:

*ne* ... **rem publicam grauet**, *honesta missione donatus uacans sibi* **otio gaudeat absolutus**. *in cuius locum posterior succedens totum numerum per certa temporum spatia* **expensis grauissimis** *releuabit.* ... *hic quoque pari liberalitate* ***otio donatus abscedat** ... *quae res non solum* **expensis grauidam rem publicam** *releuabit, sed etiam* ...

An instructive passage. The elements *res publica* - *grauare/grauis/grauidus* - *expensa*, differently combined, produce a spurious effect of variety from painfully limited verbal resources; and variation of word-order and phrasing (but not of inflexional forms) lends a similarly superficial air of novelty to **otio donatus abscedat* in 4, picking up *otio gaudeat absolutus* in 3.

[j] vii.3-6:

*quibus pro commoditate rerum circumducta et* **flexa**, *facillime ad omnes impetus parata consistat. quae quidem a fronte cochleae machina et deponitur celerius et* **erigitur** *subleuata. sed huius temo ... cita et facili conuersione* **deflexus erigitur**.

[k] vii.5 *with* xvii.1, xviii.3 *and* xix.2:

(vii.5) **in quamuis partem necessitas uocet**

(xvii.1) **quocumque utilitas uocet**

(xviii.3) **pro ut uocet utilitas**

(xix.2) **pro ut** *membrorum poscit* **utilitas**

A simple case of a repeated formula; but notice the close grouping of the last three instances.

[l] viii.1-2:

*per hunc* **facilior** *in murum* **paretur** *ascensus, ... quo* **protectior** *eadem ballista operetur; ... erit* **ergo** *huius quoque compositio uel* **fabrica** *utili et commoda inuentione* **praeparanda. ergo** *hic idem tichodifrus non* **altior** *sed* **humilior fabricatur** *ut* **intra se** *possit* **latenter** *incedentium celare uestigia: qui ...* **intra se** *tectos ab omni incursione defendit, duobus sane rotis ad promouendam machinam* **latenter** *impulsus*.

Something of a chain-reaction seems to have taken place here. The first comparative, *facilior*, persists in the writer's mind and distorts the following grammar, where the adjective *protectior* is a totally unexpected substitute for the adverb: the attraction continues into the unnecessarily comparative degree of *altior* and *humilior*. (These two words are correctly referred as adjectives to the subject, though the phrase is repeated in adverbial guise in xviii.3). There is an ineffectual attempt to conceal a poverty of vocabulary in the variation between simple and compound, *paretur... praeparanda*.

[m] ix.1 *with* xvii.3:

(ix.1) *clipeus, fixoriis* **minutis** *ad soliditatem sui diligenter* **munitus ... munit** *inferius latitantes ab incursu* **comminus uenientum**.

(xvii.3) *omnes aduersarias liburnas* **comminus uenientes**.

[n] x.1-3:

*hoc iaculi genus, quod* **in modum sagittae** *pennis* **uidetur** *instructum, ...* **in** *hostem comminus uadit: quod excipienti gemina ratione* **uidetur afferre** *perniciem: aut enim perimit* **infixum** *aut sine*

**uulnere** *dilapsum et* **in** *terram positum plantae calcantis* **infigitur**, *eo quod* **in** *partem quamlibet si fuerit conuersum,* **ex** *latere aculeo* **ex** *se eminente* **inferat uulnus. fit** *autem* **ex** *ligno* **in modum sagittae facto**, *cui ferrum subtiliter* **in** *formam uenabuli aptatum* **infigitur** ... *in summa autem parte eiusdem iaculi* **affiguntur** *pennae* ...

Consider here the six appearances of *in c. acc.* (the instance with ablative at the beginning of 3 stands outside the sequence), the unenterprising variation in the compounds of *figere*, and the (inadvertent?) repetition in *fit* ... *facto*.

[o] x.1, the author's standard introductory formula appearing with slight variations at xii.1, xiii.1, xv.2, xvi.3 and xviii.1:

(x.1) **hoc iaculi genus, quod** *in modum sagittae* **pennis uidetur instructum**, ...

(xii.1) **huius modi pugnacis uehiculi genus, quod armis** *praeter morem* **uidetur instructum**, ...

(xiii.1) *in* **hoc curriculo, quod** ... *discurrit,* ...

(xv.2) **hoc** *enim* **uestimenti genus, quod** ... *conficitur* ...

(xvi.3) **hoc** *idem* **opus** ... *ad alteram explicabitur ripam:* **quod** ... *transeundi per fluuium* ... *praebebit liberam facultatem.*

(xviii.1) **huius modi ballistae genus** ... *supra ceteras impetu et uiribus praeualere usu compertum est.*

x.1 and xii.1 share an identical verbal idiom in the relative clause. xiii.1 represents a severely adapted form of the stock phrase: and in xvi.3 the relative does not introduce a parenthesis.

[p] xi.1 *with* xviii.6:

(xi.1) *clipeos aduersarii et similiter obstantia* **ualeat penetrare**.

(xviii.6) *Danubii* ... *latitudinem* **ualeat penetrare**.

[q] xii.2 *with* xiv.1

(xii.2) **cultris** *in ordinem exstantibus* **communitur**

(xiv.1) **clipeis** *acuto ferro circumdatis* ... **communitur**

*in ordinem* in the first quotation involves a somewhat harsh case-usage after the preposition. A plausibly accusatival sense can be forcibly extracted from the phrase ('projecting into, *i.e.* so as to form, a row') but an ablative is what the more natural sense 'in a row' requires; the idea is local, not of motion or superposition. Considerations of euphony may have prompted the accusative, but a better explanation may be found in the influence of the preceding phrase, *citato cursu in pugnam rapiunt.* By contrast, *in terram positum* and *in partem quamlibet* ... *conuersum* at x.1 are, despite a context of *in c. acc.*, surprisingly accurate usages.

[r] xii.3-4 *with* xiv.4-5:

(xii.3-4) funes ... laxati *quidem* explicant, repressi *autem* erigunt falces. *qualia uero huius modi machinae* funera *hostibus immittant* ...

(xiv.4-5) *licet conuertere* laxatis funibus explicatas falces ... funis *uero, quo* laxato *aut* explicantur falces *aut* represso *super eosdem axes* eriguntur ...

[s] xv.1-4 (I quote extensively):

... *ad leuamen* corporis *armorum* ponderi *et* asperitati *subiecit. hoc enim* uestimenti *genus, quod* de *coactili ad mensuram et tutelam pectoris humani* conficitur ... *timoris sollicitudo* ... *composuit ut* hoc inducto *primum lorica uel cliuanus aut his similia fragilitatem* corporis ponderis asperitate *non laederent, membra quoque* uestientis ... *labori sufficiant. sane ne* idem thoracomachus *pluuiis uerberatus ingrauescente* pondere *adficiat* uestientem, de *libycis bene* confectis *pellibus ad instar* eiusdem thoracomachi *faciem conueniet* superinducere. hoc *igitur, ut diximus,* thoracomacho inducto, ... *soccis etiam* ... *et ferratis ocreis* inductis, superposita *galea,* ... *miles armabitur.*

[t] xvi.1 *with* xviii.8 *and* xix.2:

(xvi.1) *ne interdum necessario* itineri exercitus fluuiorum (quod plerumque euenit) occursus impediat ...

(xviii.8) *si uictricem* exercitus cursum fluuiorum *uspiam* meatus impediet ...

(xix.2) *ne sit* (quod *usu* plerumque euenit) *pro armorum latitudine siluarum densitas inaccessa* ...

[u] xvi.2-5:

... *ita ut cum idem* utres *spiritu inflati tumuerint non extollant* uterum ... desuper *autem una* parte *circulis exstantibus ex altera immittuntur uncini, et ita in formam pontis adsociata* partibus explicatur *integritas. sed hoc idem opus obliquo fluuio* ... *usque ad alteram* explicabitur *ripam* ... *funibus ualidis in medio quidem sub ipsis utribus propter* incedentium sustinendum *pondus, in marginibus autem* firmitatis *gratia* desuper *extensis* ... *super utrium compaginationem cilicia sunt* incedentium substernenda *uestigiis ne lubrica pellium confectio insistendi deneget* firmitatem ...

The final sentences of c. xvi are dominated by gerunds and gerundives, *sustinendum* - *transeundi* - *admonendi* - *substernenda* - *insistendi*; see the full context. *consistat* at the end of 5 may have been suggested, subconsciously, by *insistendi* in 4; the echo would be as insidious as the prompting of *substernenda* by *sustinendum*, both following an instance of *incedentium*. The jingle *utres* - *uterum* is comparable with the *comminus* - *comminuat* of xvii.3 (below, [v]) and is equally objectionable. *firmitas*, it may be noted, stays with the author and re-appears in xviii.10.

[v] xvii.3:

*haec eadem tamen liburna ... tanto uirium fremitu pugnam capescit ut omnes aduersarias liburnas* **comminus** *uenientes facili attritu* **comminuat**.

[w] xviii.1-11:

*huius modi ballistae genus ... supra ceteras impetu et* **uiribus praeualere** *usu compertum est: arcu etenim ferreo supra canalem quo sagitta* **exprimitur erecto** ... **ualidus** *nerui funis ... eandem sagittam magnis* **uiribus** *in hostem dimissus* **impellit**. *hunc tamen funem non manibus neque* **uiribus** *militum trahi ... fabricae ipsius magnitudo permittit ... pro difficultate rei* **uiribus** *machinis adquisitis. ballistam tamen ipsam ... cochleae machina pro ut uocet utilitas nunc* **erigit** *nunc deponit ... unius tantum otiosi ... hominis ad offerendam tantum modo* **impulsioni** *sagittam opera gubernat, uidelicet ne ... minueretur* **artis inuentio**. *ex hac igitur ballista tot et tantis* **ingenii artibus communita expressum** *telum in tantum longius uadit ... appellatione sua* **uirium** *testatur effectum. his igitur,* **inuicte** *imperator, machinarum quoque inuentionibus* **communitam** *potentiam* **inuicti** *duplicabis exercitus, hostium* **incursibus** *non ui solum et* **uiribus** *obuiando sed etiam* **artis ingenio** ... *cum ... per omnia elementa machinas reppereris* **ualituras**. ... **fugientium** *agmina persequeris ... aut si* **uictricem** *exercitus* **cursum** *fluuiorum uspiam meatus impediet, portatilis pontis peculiaris* **succurret inuentio**. *quod si nauali bello terras* **fugiens** *maria hostis obsideat, nouo* **celeritatis ingenio** ... *rotis ac bubus subacta fluctibus liburna transcurrens restituet sine mora uictoriam. quis enim huius* **uiribus** *resistet...? ...id sibi* **artis ingenio** *dotata largitur.*

The interactions of the vocabulary are horrifyingly complex, and only one of them (the rhetorical flourish *inuicte imperator ... inuicti exercitus*) can even remotely be regarded as deliberate. *firmitate* in 10 (not given above) is recollected from xvi.3 and 4, while *pro ut uocet utilitas* (3) derives from vii.5 and xvii.1 and will reappear at xix.2. *communita - communitam* is a survivor from *communitur* in xii.2 and xiv.1. Faced with the involved repetitions displayed above, it would be ungracious to draw attention to the repetition of such terms as *funis* in the early part of the chapter and *machina* throughout.

[x] xix.1-2:

*ne ... uirtus* **compressa** *minuatur* **conueniet** *ducem spatia ternis agminibus ordinare ... ne* **compressa** *populositas ... suis copiis opprimatur.* **conueniet** *tamen ... militem ... thoracomacho muniri ...*

[y] xx.1 (end) *with* xxi.1 (beginning: chapter-heading omitted):

*ut prouinciarum quies circumdata quodam* **praesidii** *cingulo inlaesa requiescat. diuina prouidentia, sacratissime imperator, domi forisque rei publicae* **praesidiis** *comparatis ...*

How are we to account for these verbal repetitions in the text? Different explanations must be invoked in different cases. Some of the recurrences are, without doubt, perfectly innocent, arising from natural re-statements of leading ideas: thus, the *anonymus* could scarcely avoid repeating the words for 'bronze' and 'gold' in cc. i-ii. Closely connected with repetitions of this type are those which result from the deficiency of Latin synonyms for a given concept - *uesicae* would be no substitute for *utres* in c. xvi - and from the ready availability of such set formulae as *huius modi*. In these cases it is not the fact that a word or an idiom is repeated which strikes us so forcibly as the remarkable incompetence shown in repeating it so often and at such close intervals. Again, some of the repetitions are a simple consequence of the abstract or technical nature of the material and the abstract terms in which the author chooses to deploy it: he is pre-occupied with such concepts as *ars, ingenium, integritas, commoditas, necessitas* and *largitio*, and insists on expressing his ideas by reference to their qualities of *facultas, difficultas, utilitas*, even - in connexion with a shaped weapon - *rotunditas*.

But it is clear that many of the repetitions originate in a sheer poverty of verbal invention. *largitio* is inevitably *profusa, aurum* and *argentum* are inextricably linked, *utilitas* and *necessitas* frequently 'summon' or 'require', successive generations 'figure' or 'mark' their coinage, the state is *grauida* or *grauata* with *expensa grauissima*; and so on. We watch with a certain wry amusement as the *anonymus* shuffles his few poor words around in a desperate attempt to make them, limited and intractable though they are, express his novel and startling perceptions; uniting, for example, in the words *ingenii artibus communita expressum* at xviii.5, four items which appear separately at different points in the surrounding context.

Yet there is something more insidious, a probably subconscious feature of the author's handling of language, to be detected in those cases where a repetition takes the form of the apparent suggestion of one word by the morphology, etymology, or even the mere sound, of a preceding one. *accendit* and *accensa* in ii.2 and 5 dictate *succendit* in 6; the group *exercitus|fluuiorum occursus* in xvi.1 (the vertical line shows the syntactical articulation of the words in context) prompts the group *exercitus cursum|fluuiorum* in xviii.8; *minutis* in ix.1 may have suggested the following *munitus*, and this certainly prompts the verb (*munit*); *incedentium sustinendum* in xvi.3 seems to be recollected in *incedentium substernenda* in the sentence next following; and *comminus* in xvii.3 may (unless the play on sound is conscious and deliberate) have suggested the verb *comminuat*, following at three words' distance. Even in cases where the author may justly be suspected of having sought variety, the effect is achieved only by the alternative use of simple and compound forms from the same root (so *paretur* ...*praeparanda* in viii.1) or by ringing the changes on available compounds (thus the sequence *infixum*...*infigitur*...*infigitur*... *affiguntur* in x.1-3).

Occasionally, indeed, the survival of a form in the author's mind seems actually to have distorted his grammar. *protectior*, standing quite unexpectedly for *protectius* (adverb) in viii.1, is preceded by *facilior*, and the combined influence of these two comparative adjectives overflows into viii.2's *non altior sed humilior* (contrast xviii.3 *non altius sed humilius*); and *in* takes accusative rather than ablative in xii.2 *cultris in ordinem exstantibus*, where an *in cum accusatiuo* immediately precedes. And the author is sometimes caught up, especially with his prepositions, in an obsessive repetitiousness which he seems helpless to control, from which he is unable or unwilling to escape, or of which, perhaps, he was not even aware; so the repeated idiom *talis...ut* in praef.13, the five *pros* in succession in the same sense at i.8-10, the *ins* and *exs* of x.1-3, *uiribus* four times repeated in xviii.1-2, and the stock introductory formula *hoc machinae (iaculi, uehiculi, uestimenti) genus, quod...*, first devised at x.1 and re-used at xii.1, xv.2, xvi.3 and xviii.1.

As a final complication, let it be suggested that the influence of words may not always have proceeded in the same, forward direction. We have no means of knowing whether, or how, or how often the author may have revised what he had written (that he made at least one revision of his work I would conjecture from the appearance of *rei publicae praesul* in an unrelieved context of *principes* in the preface; see above, section I.1.[a], my note on vocatives). It is therefore not impossible that in the course of any correction which was applied to the text certain items may have exercised a retrospective influence on preceding matter, the author rejecting, in an early part of a paragraph, a word originally written in favour of a word in some way similar to it which had suggested itself in the course of writing.

## 3 COARSENESS OF VERBAL TEXTURE

A direct consequence of the excessive limitation of the author's resources of expression and literary imagination is a perceptible coarseness of the verbal texture, the visible result of a disinclination to choose words with precision, a subservience to conventional modes of expression, and an insensitivity to the effect of words when woven into a context. Each of these factors will bear scrutiny:-

*Imprecision* of language is most evident in the case of the repeated items of vocabulary, where the fine and exact adjustment of word to meaning is replaced by the use of some favourite word whose semantic field roughly covers the desired sense. The completeness of the pontoon bridge *explicatur*, 'is achieved', by the assembly of its components; and immediately afterwards the completed bridge *explicabitur*, 'will be built out' with great ease to the opposite riverbank (xvi.2-3). At i.8, *species* means successively, at a distance of only twelve words, 'type (of coin)' and 'portrait (of a king)': it had meant 'variety' at praef.9. *utilitas*, heavily over-used by the *anonymus* as an all-purpose abstract - its inflexion occur, in total, fifteen times in a text of some 3340 words - means 'public service' (praef.2, *bis*), 'expedient' (praef.2), 'discovery' or 'progress' (praef.6), 'welfare' (iv.1), 'requirements' (xix.2), even - unexpectedly - 'utility' (vii.2). But there is no cause for alarm in the author's extended use of *ars*, which now approaches 'art' (praef.6) and now 'science' (xviii.11), taking in on the way the meanings 'skill' (xiii.1) and 'technique' (xvi.2); these are all senses which the word may legitimately bear.

*Conventionality* of language appears most clearly in the threadbare adjective and the trite metaphor. The *anonymus'* epithets - and his prose is not distinguished by a profusion of adjectives - are for the most part flat and sometimes merely decorative. The *res publica* is inevitably *felix* (praef.1) and so are its *limites* (xviii.11) - this, despite the real state of affairs which our author perceives all too clearly. *fides* is *uera* at i.10 and *integra* (this is perhaps suggested by the earlier *integritatem*) at iii.3: the adjectives are ornamental to the point of tautology. Or should one rather suppose that these are symptoms of a linguistic inflation, parallel with the inflation of the late imperial currency, with two debased words acting where one would formerly have sufficed? As for metaphors, the *anonymus* cannot resist the venerable image of medicine and cure (xvi.1, xxi.1), and a machine is predictably *endowed* with all the refinements of art (xviii.11), while hatreds are naturally *inflamed* at ii.6. Mercifully the emperor is never discovered at the helm of the ship of state, though he is asked to *enlighten* a situation at xxi.1: the verb in this last image, *illustrare*, is used in a different but equally tired sense at praef.7, more boldly in praef.10.

*Insensitivity* towards the handling of words manifests itself in many different ways - the redundant expression, the badly-managed metaphor, the unpremeditated ambiguity, the harsh or elliptical turn, the cacophonous phrase, the ill-ordered construction, the awkwardness of transition, the sudden lapse of tone, the clash of the images which words contain. [1] Redundancy of expression has been partly touched on in my remarks on conventionality of epithets, above; more extensive examples of the same regrettable tendency towards pleonasm can be seen in xix.5 *duorum gestatus ministerio uirorum* and xxi.1 *remedium ad ciuilium curarum medicinam*. [2] I have also referred above to the general triteness of the *anonymus'* metaphors; but their very conventionality begins to seem a virtue when we reflect that, where the writer does strike out an extended image for himself, the result is an embarrassing failure, inconsistent, tautologous and dissonant (ii.1 *huius auaritiae origo hinc creditur emanasse*); though he may be considered to have been more successful with the one-word image in *cingulum*, applied to a wall with mile-castles, and apologised for with a *quodam*, just before the end of xx.1. [3] Ambiguity is one of the most powerful weapons at the command of the literary artist, but, appearing accidentally, it may suggest a certain indifference to linguistic finesse. Ambiguities - no doubt unintentional - can be detected here and there in the *anonymus'* language: in praef.8, is *romano nomini* coupled more closely with *debitos* or *affectus*? In vi.1, should one take *ubique* with the preceding participle or the following verb? In xvi.1, is *interdum* with the adjacent *necessario* or the remote *impediat*? And in xvi.2, does *partibus* depend on *adsociata* or on *explicatur*? The explicitly syntactical ambiguity surrounding the cases of *quicquam* and *machina* in xiii.1 has been handled earlier (under the heading *Accusative* in section I.1.[a] above); it remains to discuss the problem of xviii.5 *in tantum longius uadit*, where the simultaneous presence of two constructions blurs the syntactic articulation. At first view, *in tantum longius* appears to cohere as an intensified *ita longe*; and a feeling that this construction is present persists even when it is seen that *longius* in fact applies less closely to the preceding adverbial group than to the verb, and that the resulting verbal complex *longius uadit* is modified by *in tantum* replacing *ita* as the direct antecedent of the consecutive clause. [4] A harshness can be detected in the use of *uarie* to mean both 'in various ways' and 'for various reasons' in ix.1; in the undue compression of meaning into the solitary *diuturnior* in i.8; and in *suggerenda* without an indirect personal object in pr.1: while an ellipticality of expression produces the jarring *nullo uetustatis more* of i.2 and the badly-coordinated *funibus ualidis in medio quidem sub ipsis utribus propter incedentium sustinendum pondus, in marginibus autem firmitatis gratia desuper extensis* in xvi.3. [5] The careless juxtaposition of the author's beloved abstracts in *-tas* often wounds the ear; iii.1 displays a total failure to control these and other sounds, *ementis enim eundem solidum fraudulenta calliditas et uendentis damnosa necessitas difficultatem quandam ipsis contractibus intulerunt, ne rebus possit interesse simplicitas*. [6] The double syntax of ii.8 *famosa regna considera, qua haec honoris laude...frugalitas incorrupta commendet* is very strained, and xvi.2 almost disintegrates through the intrusion, shortly after the beginning, of a long parenthesis. [7] The transition to a relative clause is very badly managed in i.8, where the *quae* of the final sentence refers back to *speciem* across two neuters which at first it seems to pick up; and the occasionally abrupt transitions of thought are, in general, ineffectively disguised by connective particles (above, section I.5.[b]). [8] The uncertainty of tone of the *anonymus*, his failure to sustain a consistent level of diction and style, can be seen in miniature in pr.15, *magnum uobis munus concessu diuinitatis apporto, asserens prouidentia pietatis uestrae armorum uigorem et cunctam rem publicam praedictis remediis subleuandam*, where the pedestrian *praedictis* brings the flight of eloquence to the ground. On the larger scale, formal words such as *opitulante* (pr.7), *fas* (pr.4), *nuncupamus* (ii.9; *nuncupata* with the unique by-form *fulminalis* at xviii.6), *gestiui* (pr.16), *cita* (vii.5) and *exordium* (viii.1) coexist with such flat or colloquial terms as the *bene* with adjectives at xi.1 and xv.3, the diffident *quodam modo* of xvii.1 and xviii.9, the cliché *huius modi* at xix.5, *mamillata* (xi chapter-heading), *tractantibus* (pr.9 *bis*) and *rebus* in iii.1 and iv.2 (*bis*). It was of the nature of the subject and its treatment that appeals to such abstract attributes of sovereignty as *clementia, felicitas, maiestas* and *serenitas* should consist with the names of utilitarian items of military equipment: but the uncompromising use of technical terms almost *passim* - fiscal in cc. i-iii, administrative in c. iv, military in cc. v-xx - contrasts with the painful literalness of exposition of *parma, hoc est modicus clipeus* in ix.1 and *thoracomacho inducto (qui graeca appellatione ex tuitione corporis nomen assumpsit), soccis etiam (hoc est calciamentis) et ferratis ocreis inductis* in xv.4. And [9] the horrible clash of images in *picturam... coloribus adumbratam* in vi.5 results from a failure to think closely enough about words; the *anonymus* has simply not perceived the metaphor which *adumbrare* scarcely conceals.

## 4 THE STYLE: A GENERAL SUMMARY

The *anonymus* has composed an *oratio* (praef.17; vi.5), at once an address, a speech, a supplication, a request: and his prose constantly threatens to break down under the strain which the varied demands of the material and treatment impose upon it. Plain, compressed and direct in the specification of a machine, it aspires to eloquent commendations charged with the windy inflation of rhetoric; dispassionate in discussion, it will suddenly surge up, driven by the force of conviction, in an outburst of febrile energy; fiercely independent in outlook, it preserved the frigid courtesies of an address to authority; closely analysing cause and effect, it fails to choose words accurately; clogged with abstractions, it seeks to convey remedies for a frightful reality; pervaded by verbal repetition, it conveys extreme originality of thought. Rebarbative, coarse, uneven and difficult, it is not wholly devoid of a certain touching appeal; the Greek names for the inventions, clumsy though they are, are irresistibly endearing; and I would ask the reader wearied with rhetoric and mechanics to consider sympathetically a sentence of great charm, xv.4, where the *anonymus*, having stumbled though an etymological note and a *hoc est* gloss, builds up a sequence of short clauses into a climax of pace and attack, and the chapter ends, like the most perfect of logical demonstrations, with an air of triumphant delight.

Samuel Johnson said of Richard Bentley's only known English poem that 'they are the verses of a man of strong mind, but not accustomed to write verse; for there is some uncouthness in the expression'. Slightly adapted, this verdict will describe admirably the prose style of the author of *de rebus bellicis*. And can one go further? Is this uncouth and forcible prose the work of a man of strong mind not merely not accustomed to write, but actually not accustomed to write Latin? I do not suggest that the *anonymus* was a Greek: a Greek would have handled the compound machine-names more fastidiously: but the stuttering language of the treatise, the author's intimate acquaintance with the army and the enemies of the empire, his vivid perception of economic failure (so accurate that it best befits an external, not an internal observer), and the possible implication of a strange sentence in the preface (7: *nam...barbarae gentes...minime...a rerum inuentione...habentur alieni*) – all these indications, put together, suggest to me that the *anonymus* was a man of large intellectual curiosity with an elementary grounding in rhetoric but little experience of formal written expression; that he had acquired, probably in the provinces, some experience of financial affairs and the fiscal system; that he was an army man, perhaps an officer, perhaps connected with the artillery, but now in retirement (praef.16 *otio persuasus, non adeo a rerum commoditatibus peregrinus*); and that he was perhaps himself, by origin, a member of one of the *barbarae gentes* who were howling around the Roman empire, and whose assaults his little treatise, in all its pathetic urgency, was in great part designed to resist.

# 6 INDICES

Concordance to the Latin text

Analysis of the contents of this edition

## Concordance to the Latin text

13 a: pr.4,7,15,16; ii.3; iii.2; vi.5; vii.4; xii.2; xiii.1 *bis*; xiv.1,3

8 ab: iv.1,3; v.5; viii.2; ix.1; xiv.4; xix.6; xxi.2

1 abscedat: v.4

1 absolutus: v.3

1 abstinere: ii.8

1 abundans: xix.1

1 abunde: pr.3

2 ac: vi.2; xi.1

1 accedat: v.4

1 accedit: iv.1

1 accendit: ii.2

1 accensa: ii.5

1 accidere: pr.7

1 accipiet: xi.1

1 accommodata: pr.3

2 aciem: pr.13; xix.8

1 acquirendis: pr.11

1 acquisita: pr.6

1 aculei: x.2

1 aculeo: x.1

1 acumen: xi.1

1 acutis: xix.6

1 acutissimae: xii.3

1 acuto: xiv.1

36 ad: pr.5; i.5,8,9; ii.2,6; iv.1; v.1,3,4,5,6; vii.2,3; viii.2; ix.1; xiii.1 *bis*; xiv.1,4,5; xv.1 *bis*,2,3; xvi.1,3; xvii.1; xviii.3,4,11; xix.2,3,6,9; xxi.1

1 adeo: pr.16

1 adficiat: xv.3

2 adhaerentes: x.2; xvii.2

1 adhaerentibus: xi.1

1 adhibenda: iii.2

1 adhuc: v.6

1 adiuncti: xvii.2

2 adiuta: xi.1; xv.2

1 admiratio: xiv.1

1 admonendi: xvi.4

1 adnectenda: pr.11

1 adnexis: xvi.2

1 adnitentes: xviii.2

1 adquisitis: xviii.2

1 adsociata: xvi.2

1 adueniant: xix.6

1 aduersarias: xvii.3

1 aduersarii: xi.1

1 adumbratam: xi.5

1 aequalitate: xvi.2

1 aerarii: i.1

1 aere: ii.1

## Analysis of the contents of this edition

Contents of *editio princeps*, 62-63; their order, 63; omission of *de gradibus cognationum*, 63-64. *ex ultimis Britannis*; Moelbrigte; the origin of the *spirensis*, 65-66.

NOTES

*Substantive notes*: on *codices descripti*, 66 (n.2): entries in the Speyer minute-book relating to the *spirensis*, M, and Ott Heinrich, 67-68 (n.4): the copy for Ghelen's Ammianus, 68-69 (n.13): Ghelen's misidentification of the *imperator* and *filii* addressed in *d.r.b.*, 69-70 (n.17): topographical texts at Speyer, 70 (n.25): the *altercatio Hadriani et Epicteti* as a component of the *notitia*-group, 70-71 (n.26): composition and date of the *spirensis*, 71 (n.29); its presence at Speyer 1436-1552, 72 (n.39): nature of copy for *editio princeps* of *notitia*-group, 73 (n.42): copy for Bilde's Velleius, Pickel's Hrotsuith, the Subiaco Augustine, 73 (n.44): collation of the 1552 *notitia* volume, 74 (n.45); a problem in its last quire perhaps reflecting on contents, 74 (n.45): official texts in insular scriptoria, Alcuin and the *altercatio*, 74-75 (n.48).

APPENDIX: PREVIOUS EDITIONS OF THE TEXT

Ghelen, Schrijver, 76: Panciroli, Labbe, Schneider, Reinach, 77: Thompson, 78.

EXCURSUS: THE PRINTING OF THE *EDITIO PRINCEPS*

Two dated illustration-blocks, 79-81: time-scale of block-cutting and availability of exemplar, 82. Continuity of printing, 82: demonstrated from wrong-fount and broken letters, 83-84; from repeated illustration-blocks, 84-89; from blind impressions of type packing out chase, 89-91. Conclusions: the *spirensis* as exemplar of the *editio princeps*, 91-92.

## TEXTUAL CRITICISM

*Substantive notes*: a correction and supplements in pr.3, 95-96: a pre-archetypal reading in pr.5, 96: false concord retained in pr.7, 96: Ghelen with PV (and so Σ?) against CM in pr.9, 97: textual disturbance in pr.10, 97: a crux in i.1, 98-99: a supplement in i.3, 99: a lacuna detected in i.6, 99-100: insular error of V in ii.6, 100: supplements in ii.6, 100: a correction in v.4, 101: dittography in Σ in v.7, 102: M V and Ghelen in difficulty with a Σ reading in vii.6, 102-103: Ghelen again with PV (and so Σ?) against CM in a reading decided by clausula in ix.1, 103: variants and a deletion in x picture-caption and chapter-heading, and a theory of titling, 103-104: a supplement in xi.1, 104:

## LANGUAGE AND STYLE

### GRAMMAR



### VOCABULARY AND COMPOSITION



### STYLISTIC DEVICES AND CHARACTERISTICS

# 7 MANUSCRIPT ILLUSTRATIONS

## THE ILLUSTRATIONS

In the following plates the twelve illustrations of *de rebus bellicis* are reproduced as they appear in the following manuscript sources: C (*oxoniensis canonicianus lat. misc.* 378); M (*monacensis latinus* 10291 in text); M* (the set of illustrations traced from the *spirensis* and bound up at the back of M); and P (*parisinus latinus* 9661). We are most grateful to the Bibliothèque Nationale, Paris; the Bayerische Staatsbibliothek, Munich; and the Bodleian Library, Oxford, for supplying the photographs used in preparing these plates and for allowing us to publish them.

PLATE I: COMMODAE AVCTORITATIS VARIAE PRISCORVM MONETAE

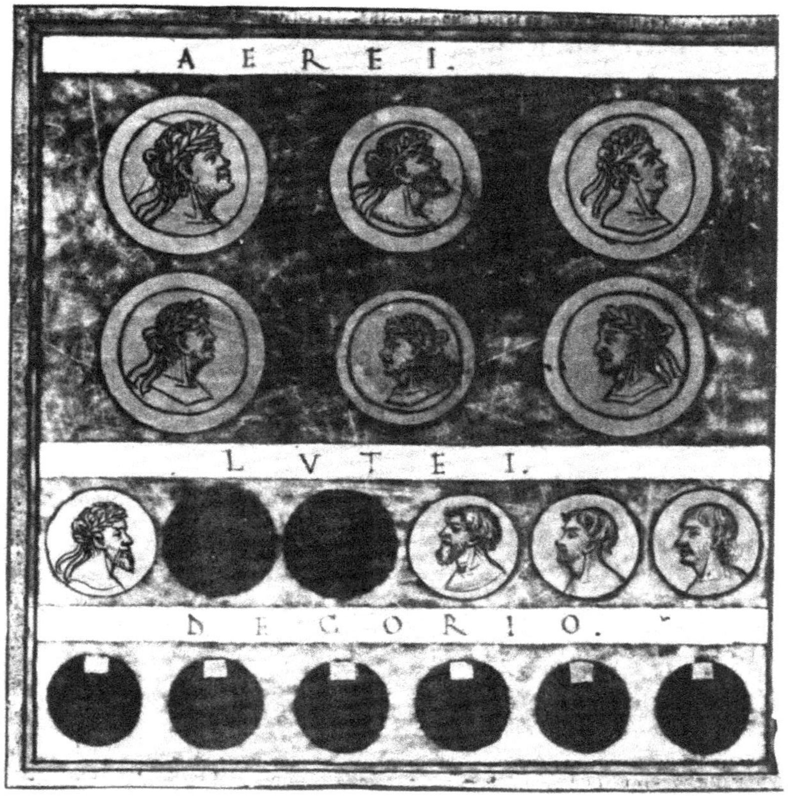


(a) as in C

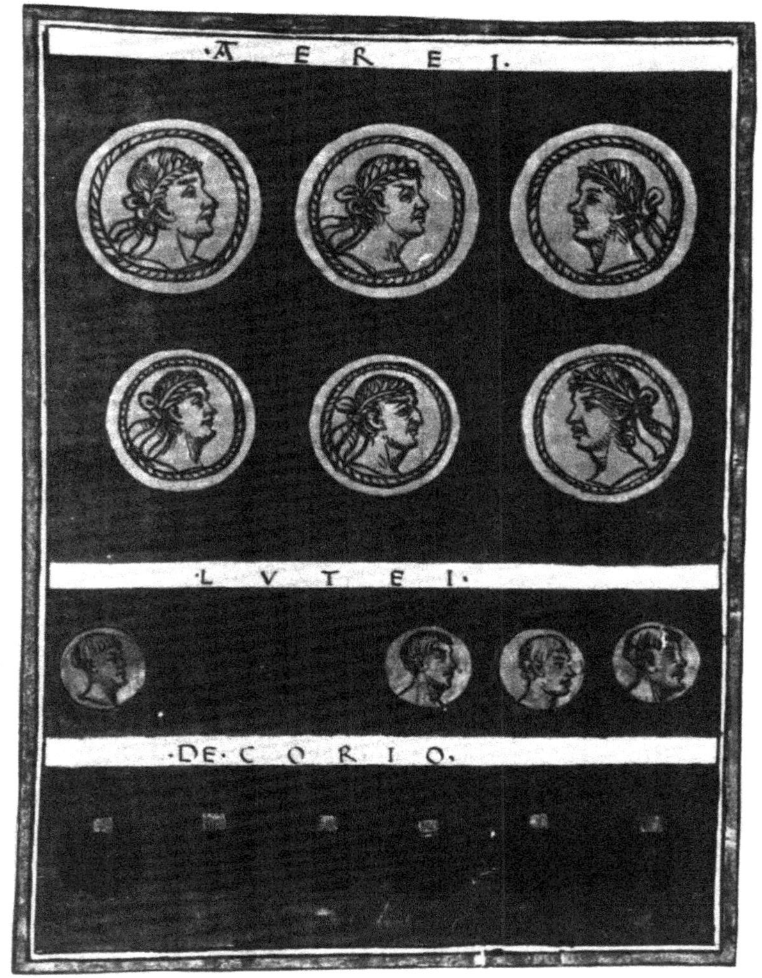


(b) as in P

PLATE II: COMMODAE AVCTORITATIS VARIAE PRISCORVM MONETAE

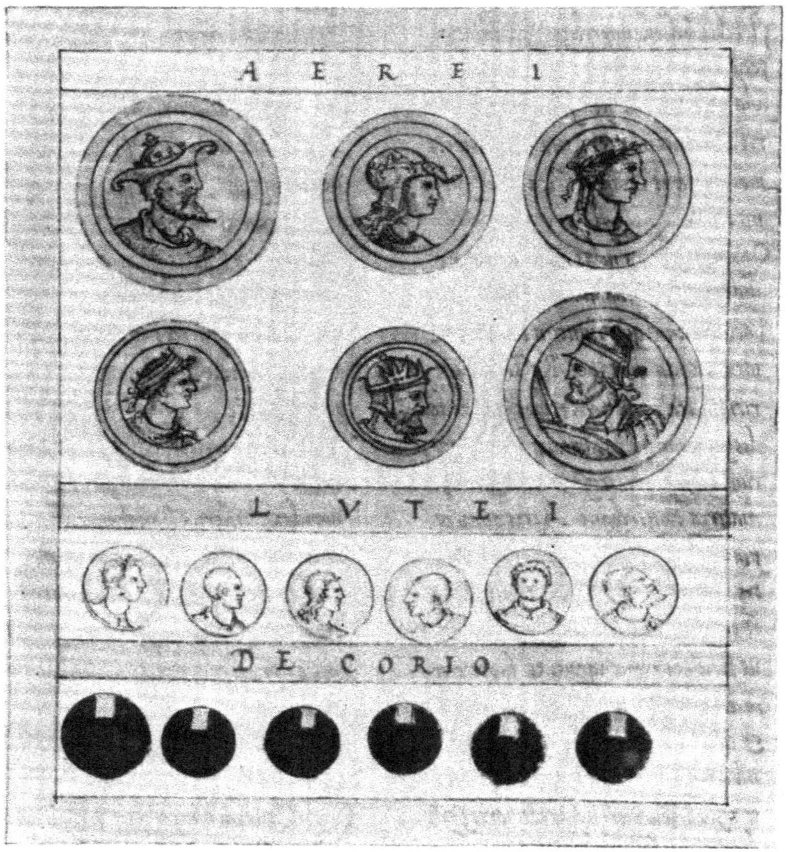


(a) as in M

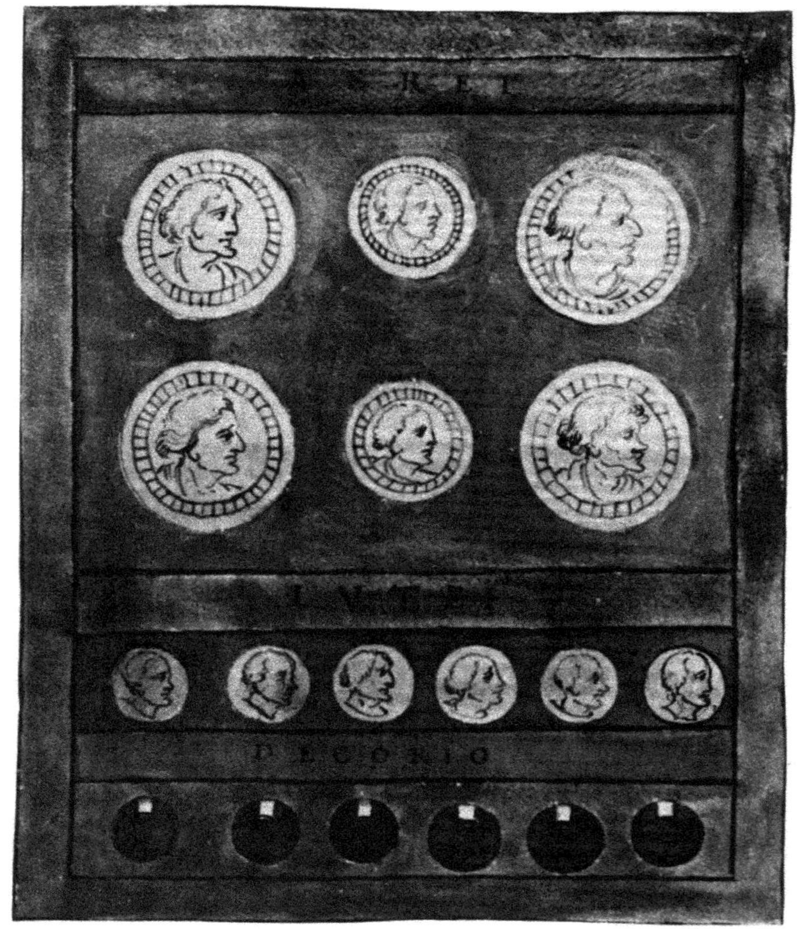

(b) as in M*

PLATE III: FELIX INCHOATIO SACRAE DIVINAEQVE MONETAE

(a) as in C

(b) as in P

PLATE IV: FELIX INCHOATIO SACRAE DIVINAEQVE MONETAE

(a) as in **M**

(b) as in **M***

PLATE V: BALLISTA QVADRIROTIS

(a) as in C

(b) as in P

PLATE VI: BALLISTA QVADRIROTIS

(a) as in M

(b) as in M*

PLATE VII: TICHODIFRVS CLIPEOCENTRVS

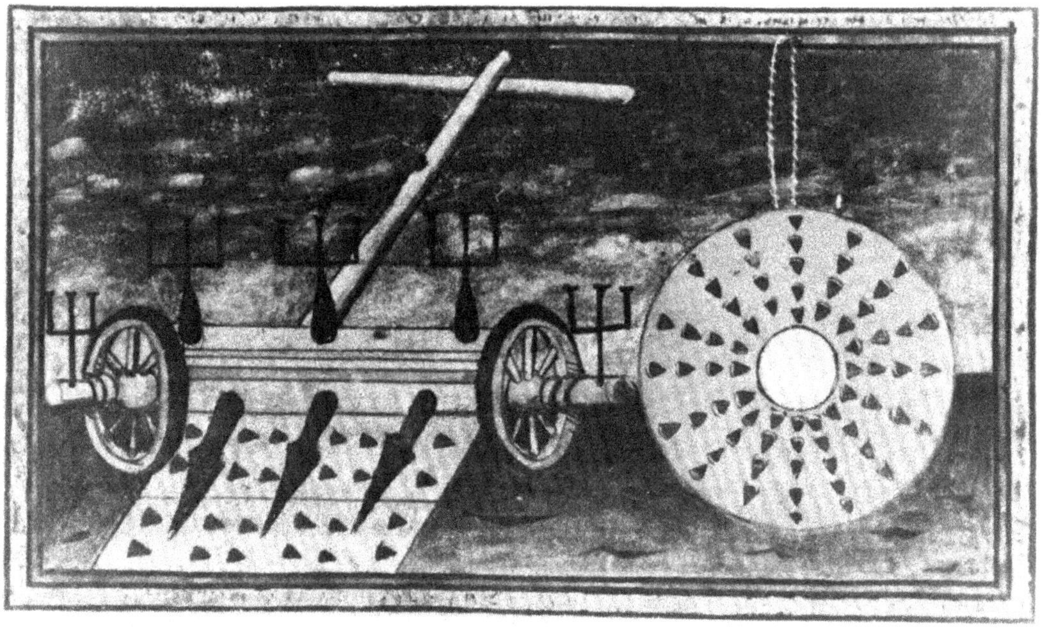

(a) as in **C**

(b) as in **P**

PLATE VIII: TICHODIFRVS CLIPEOCENTRVS

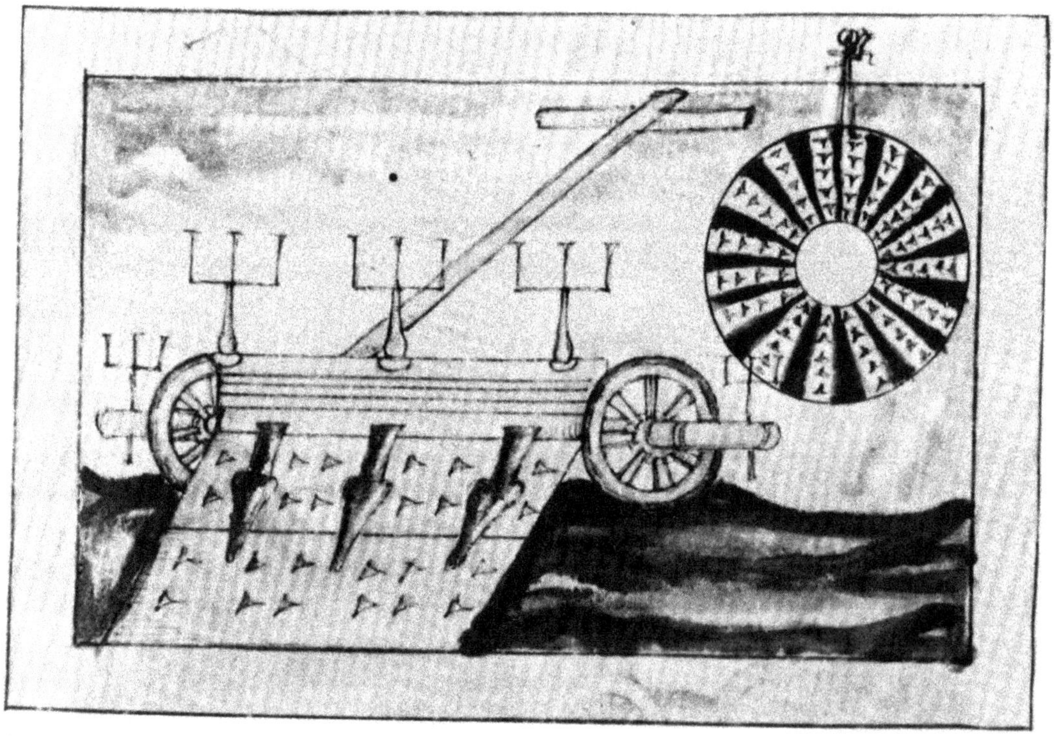

(a) as in M

(b) as in M*

PLATE IX: PLVMBATA [ET] TRIBOLATA  PLVMBATA MAMILLATA

(a) as in **C**

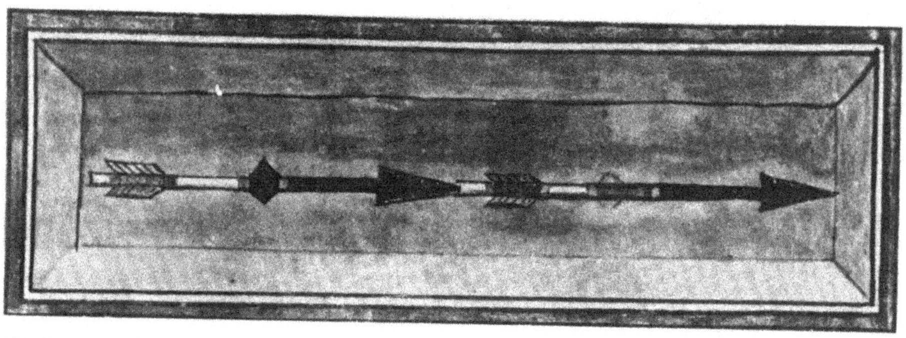

(b) as in **P**

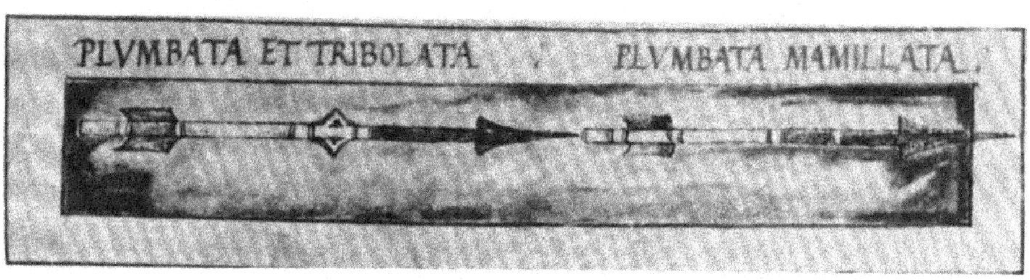


(c) as in **M**

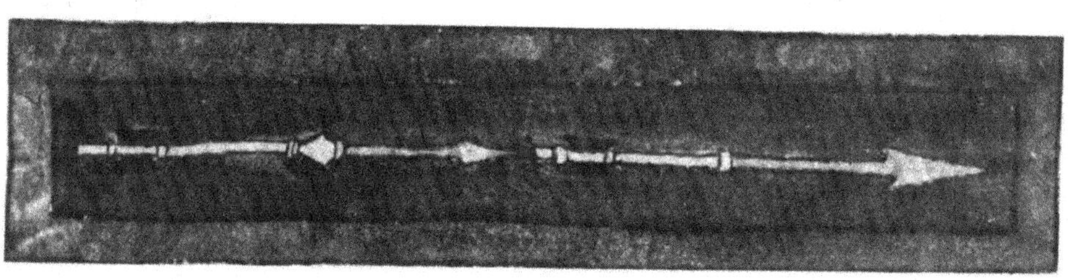

(d) as in **M***

PLATE X: CVRRODREPANVS

(a) as in C

(b) as in P

PLATE XI: CVRRODREPANVS

(a) as in **M**

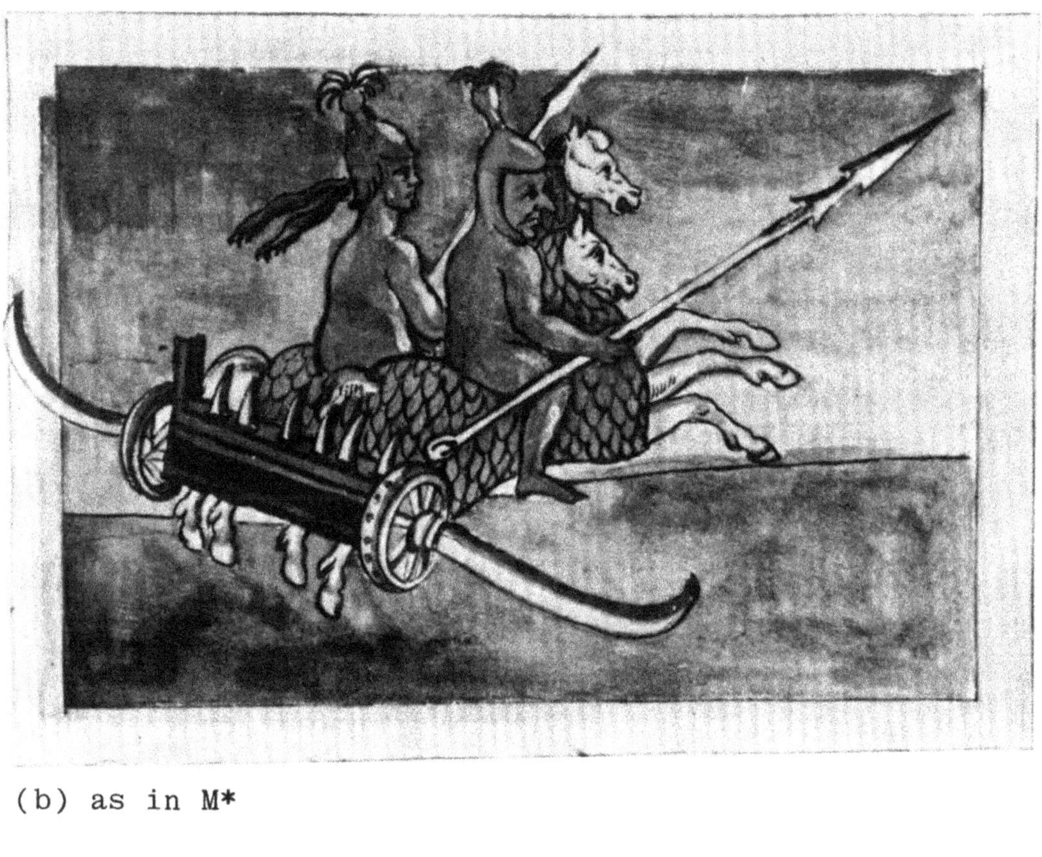

(b) as in **M***

PLATE XII: CVRRODREPANVS SINGVLARIS

(a) as in C

(b) as in P

PLATE XIII: CVRRODREPANVS SINGVLARIS

(a) as in M

(b) as in M*

PLATE XIV: CVRRODREPANVS CLIPEATVS

(a) as in C

(b) as in P

PLATE XV: CVRRODREPANVS CLIPEATVS

(a) as in **M**

(b) as in **M***

PLATE XVI: THORACOMACHVS

(a) as in C

(b) as in P

PLATE XVII: THORACOMACHVS

(a) as in **M**

(b) as in **M***

PLATE XX: LIBVRNA

(a) as in **C**

(b) as in **P**

PLATE XIX: ASCOGEFYRVS

(a) as in M

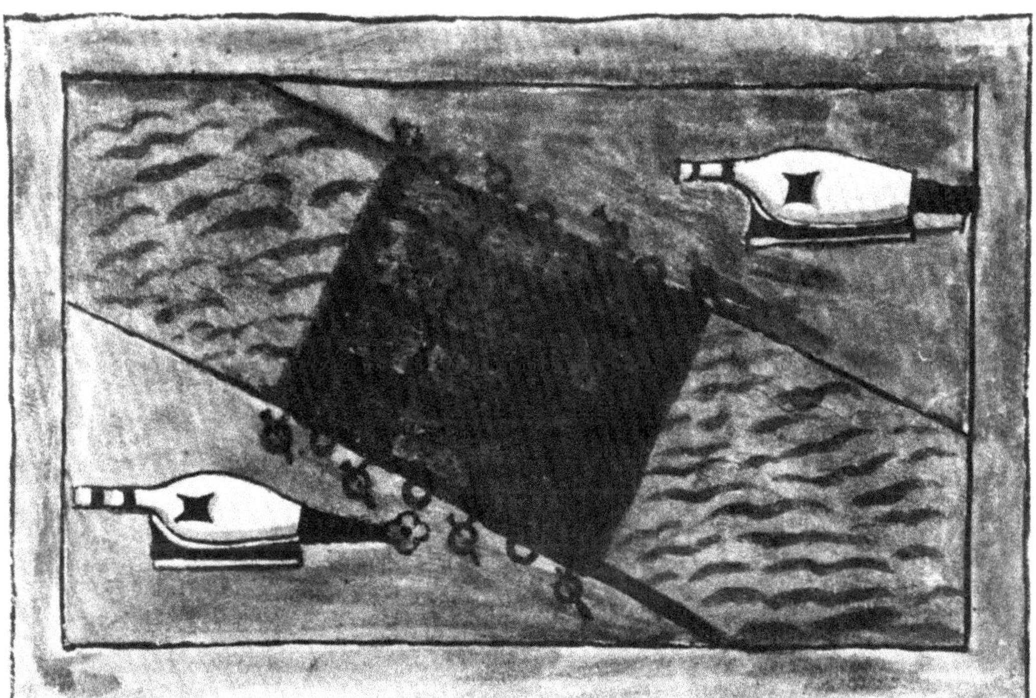

(b) as in M*

## PLATE XVIII: ASCOGEFYRVS.

(a) as in **C**

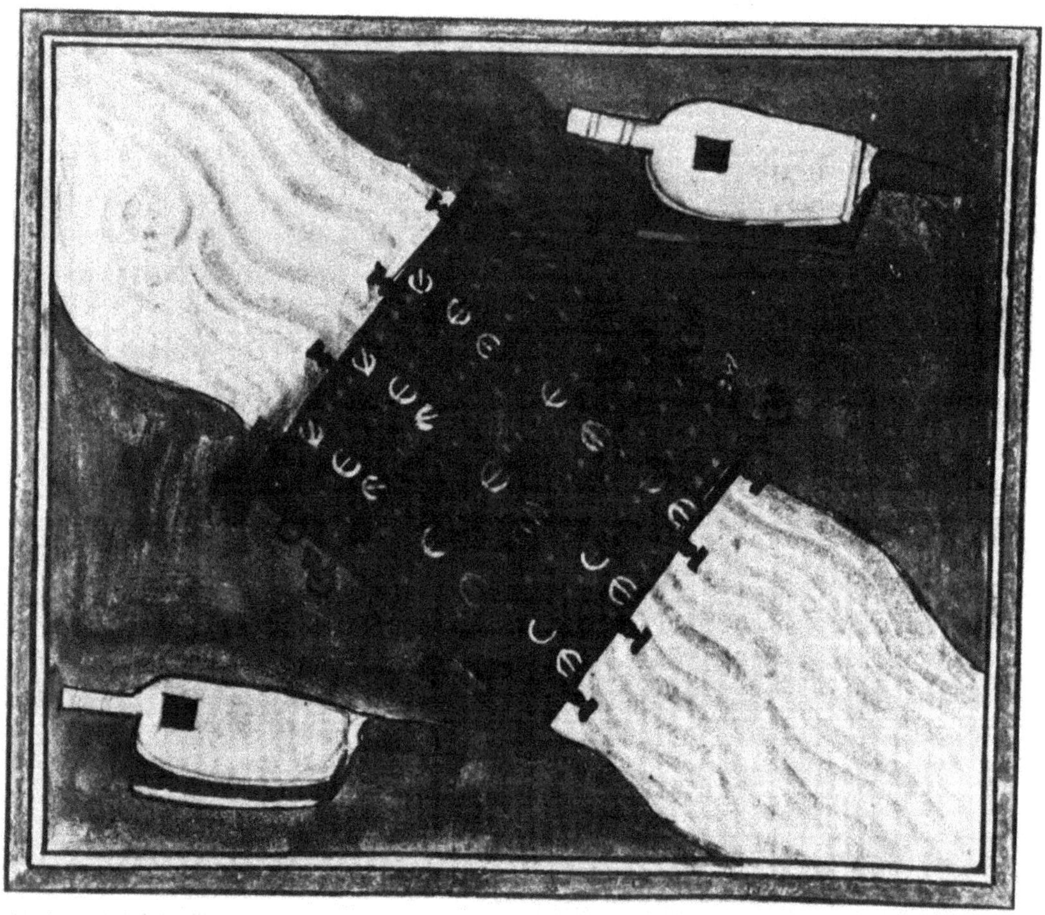

(b) as in **P**

PLATE XXI: LIBVRNA

(a) as in **M**

(b) as in **M***

PLATE XXII: BALLISTA FVLMINALIS

(a) as in C

(b) as in P

PLATE XXIII: BALLISTA FVLMINALIS

(a) as in **M**

(b) as in **M***

www.ingramcontent.com/pod-product-compliance
Lightning Source LLC
La Vergne TN
LVHW071203210926
848482LV00015B/714
* 9 7 8 0 8 6 0 5 4 0 6 3 2 *